THE PALGRAVE CONCISE
HISTORICAL ATLAS
OF
EASTERN EUROPE

ALSO BY DENNIS P. HUPCHICK AND HAROLD E. COX

The Palgrave Concise Historical Atlas of the Balkans

THE PALGRAVE CONCISE
HISTORICAL ATLAS
OF
EASTERN EUROPE

DENNIS P. HUPCHICK *and*
HAROLD E. COX

palgrave

PALGRAVE CONCISE HISTORICAL ATLAS OF EASTERN EUROPE
Copyright © Dennis P. Hupchick and Harold E. Cox, 2001.

First published 2001 by **PALGRAVE™**
175 Fifth Avenue, New York, N.Y. 10010 and
Houndmills, Basingstoke, Hampshire RG21 6XS.
Companies and representatives throughout the world.

PALGRAVE™ is the new global publishing imprint of St. Martin's Press LLC
Scholarly and Reference Division and Palgrave Publishers Ltd
(formerly Macmillan Press Ltd).

ISBN 0-312-23984-X hardback
ISBN 0-312-23985-8 paperback

Library of Congress Cataloging-in-Publication Data
available from the Library of Congress.

A catalogue record of this book is available from the British Library.

Original concept, base maps, and texts by Dennis P. Hupchick.
Finished maps and text graphics by Harold E. Cox.

First edition: September 2001
10 9 8 7 6 5 4 3 2

Printed in the United States of America.

Contents

Preface

The positive reception given our original *A Concise Historical Atlas of Eastern Europe* (St. Martin's, 1996) has been highly gratifying. We were pleasantly surprised to discover that there existed a serious interest in our publishing a second, expanded edition of that work encompassing the events that have occurred in the Balkans since 1991. In doing so, we decided to take the opportunity to refine or make minor revisions in some of the maps and texts of the first edition. This new edition includes two new maps dealing with the post-1991 Bosnian and Kosovo crises, eleven revised first-edition maps, and an updated selected bibliography.

For those readers unfamiliar with the earlier edition, it should be noted that the idea for the concise atlas arose in a 1993 conversation between ourselves concerning the lack of commercially available classroom maps specifically designed for introductory courses in East European history. We originally thought of creating a set of maps for use by students at our own university but, since it seemed likely that instructors of East European history at other institutions were dealing with the same problem, we decided that it would be beneficial to produce them for general publication.

The primary purpose of this concise atlas is to provide students and interested general readers with a basic, affordable visual aid for grasping the geopolitical situation at selected important moments in the history of Eastern Europe. It does not attempt, nor is it intended, to offer a comprehensive overview of every aspect of Eastern Europe's history. Thus, there are no specialized maps, tables, or charts dealing with economic patterns, urbanization, vegetation, land use, annual rainfall, ecclesiastical jurisdictions, religious movements, education, transportation, industrial development, demographic movements, and other such topics. Nor does it pretend to offer *definitive* cartographic representations of the periods and events covered. It also should be understood that a work of this kind can sometimes unintentionally be misleading to the historical neophyte. Geopolitical maps require the presentation of states bounded by borders, but hard and fast state "borders," as we know them today, did not evolve until the late eighteenth century. The reader, therefore, should bear in mind that the state borders appearing in the maps prior to that time are intended to provide an approximation of the territories controlled in some form or another by the various state authorities, and that the authorities' territorial control within those states may have ranged from direct to nominal at any given time.

The maps are rendered in two- rather than full-process color. Only those elements deemed necessary for a general understanding of the topics presented are included. Most rivers and mountain ranges either do not appear or do so relative to their informational purpose within any given map. Our decisions concerning scope and presentation were based on considerations of the work's fundamental purpose—basic geopolitical information—and cost—affordability.

Each map is accompanied by a page of text. Again, the individual texts are intended to provide a broad perspective on the particular periods or issues represented in the maps. They are not meant to be mere descriptions of specific map elements. Since this concise atlas will be used best as a supplemental resource by the student or general reader, the texts do not present a truly comprehensive history of Eastern Europe. Numerous factual gaps and lapses exist within and among the texts. Likewise, space limitations make it impossible to provide explanations for every foreign or specialized term used in the texts. In both cases, it is assumed that such information will be available to the user from sources outside of this publication.

Regarding spelling, most foreign common terms and proper names appearing in the atlas are rendered in or near their native spellings. Exceptions to this approach are: (1) terms generally known to English-speakers in their Anglicized forms (such as the names of states, certain cities, and geographic elements); and (2) the first names of Greek, Russian, and German individuals. Place names generally are given in their contemporary forms. Although some scholars may take issue with this decision, it should be stated that, in cases other than blatantly ahistorical instances (such as calling Constantinople "Istanbul" or Adrianople "Edirne" prior to their Ottoman conquest), changing place names, while technically accurate, has little import for anyone other than specialists. For the student and general readership to whom this atlas is addressed, such name changes tend to be more confusing than informational. Turkish terms are spelled in the Latin characters currently used in Turkey with the appropriate diacritical marks.

Two approaches are taken to transliterating into Latinized form Slavic terms that natively are written in the Cyrillic alphabet. A "phonetical" system is used for Bulgarian and Russian (for example: ч is rendered as **ch**; ш as **sh**; ц as **ts**; й as **i**; ж as **zh**; я as **ya**, etc.). In the cases of Serbian and Macedonian (Cyrillic languages in pre-1991 Yugoslavia), a "linguistic" system using diacritical marks with some characters is employed (for example: ч is rendered as **č**; ђ as **ć**; ш as **š**; ц as **c**; й as **j**; ж as **ž**; я as **ja**, etc.), which is based on the Latin, "Croat" form of Serbo-Croatian commonly used in the West for transliterating "Yugoslav" languages.

We wish to thank Michael Flamini, vice president and editorial director at Palgrave, his assistant Amanda Johnson and her editorial staff, and Alan Bradshaw, along with his production staff, for their thoughtful and creative input and overall support of our efforts.

Dennis P. Hupchick
Harold E. Cox
Wilkes-Barre, PA

General Key to the Maps

Color is employed in this atlas as a tool to make it more "user-friendly" than if it were printed in black and white. On the maps, colored lines and shaded areas highlight important geopolitical developments and help to simplify complex situations that might otherwise be confusing for the user.

International borders	– – – – – – – – –
Regional boundaries	– – – – – – – – –
Names of states	**YUGOSLAVIA**
Names of regions	*MACEDONIA*
Names of ethnic groups	*Ruthenians*
Names of rivers	*Danube R.*
Names of cities	Vienna

Introductory Maps

Map 1: Eastern Europe — Political, 2001

ALBANIA
Area in sq. mi. (sq. km.): 11,097 (28,489)
Population: 3,334,000
Ethnicity (%): Albanian (90); Greek, Vlah, Gypsy, Bulgarian (10)
Languages: Albanian; Greek
Religions (%): Muslim (70); Albanian Orthodox (20); Roman Catholic (10)
Type of Government: Parliamentary democracy

BOSNIA-HERCEGOVINA
Area in sq. mi. (sq. km.): 19,776 (51,233)
Population: 4,618,800 (prior to 1992-95 war)
Ethnicity (%): Bosniak/Muslim (44); Serb (31); Croat (17); other (8)
Language: Serbo-Croatian
Religions (%): Muslim (40); Orthodox (31); Roman Catholic (15); Protestant (4); other (10)
Type of Government: Federated parliamentary democracy

BULGARIA
Area in sq. mi. (sq. km.): 42,855 (110,994)
Population: 8,832,000
Ethnicity (%): Bulgarian (85); Turk (9); Macedonian, Gypsy, other (6)
Languages: Bulgarian; Turkish; other (as ethnicity)
Religions (%): Bulgarian Orthodox (85); Muslim (13); Roman Catholic, Jewish, Protestant, other (2)
Type of Government: Republic

CROATIA
Area in sq. mi. (sq. km.): 21,824 (56,538)
Population: 4,694,400
Ethnicity (%): Croat (78); Serb (12); Magyar, Slovene, other (10)
Languages: Albanian; Greek
Religions (%): Roman Catholic (77); Orthodox (11); Muslim (1); Protestant (1); other (10)
Type of Government: Parliamentary democracy

CZECH REPUBLIC
Area in sq. mi. (sq. km.): 30,379 (78,703)
Population: 10,389,000
Ethnicity (%): Czech (94); Slovak (3); Polish, German, Gypsy, Magyar, other (3)
Languages: Czech, other (as ethnicity)
Religions (%): Roman Catholic (40); atheist (40); Protestant (5); Orthodox (3); other (12)
Type of Government: Parliamentary democracy

GREECE
Area in sq. mi. (sq. km.): 50,962 (131,990)
Population: 10,010,000
Ethnicity (%): Greek (93); Macedonian (2); Turk (1); other (4)
Languages: Greek
Religions (%): Greek Orthodox (97), Muslim (1); other (2)
Type of Government: Republic

HUNGARY
Area in sq. mi. (sq. km.): 35,900 (92,980)
Population: 10,324,000
Ethnicity (%): Magyar (90); Gypsy (4); German (3); Serb (2); other (1)
Language: Magyar
Religions (%): Roman Catholic (67); Calvinist (20); Lutheran (5); Jewish, atheist, other (8)
Type of Government: Parliamentary democracy

MACEDONIA
Area in sq. mi. (sq. km.): 9,778 (25,333)
Population: 2,194,000
Ethnicity (%): Macedonian (67); Albanian (21); Turk (4); Serb (2); other (6)
Languages: Macedonian; Albanian; Turkish; Serbian; other
Religions (%): Orthodox (59); Muslim (26); Roman Catholic (4); Protestant (1); other (10)
Type of Government: Parliamentary democracy

POLAND
Area in sq. mi. (sq. km.): 120,700 (312,612)
Population: 38,520,000
Ethnicity (%): Polish (98); Ukrainian, Belorussian, other (2)
Language: Polish
Religions (%): Roman Catholic (95); Uniate, Orthodox, Protestant, and Jewish (5)
Type of Government: Parliamentary democracy

ROMANIA
Area in sq. mi. (sq. km.): 91,699 (237,499)
Population: 23,172,000
Ethnicity (%): Romanian (89); Magyar (9); German, Turk, Russian, Ukrainian, Gypsy, other (2)
Languages: Romanian; Magyar; German
Religions (%): Romanian Orthodox (70); Roman Catholic (6); Protestant, Jewish, other (6)
Type of Government: Democratic republic

SLOVAKIA
Area in sq. mi. (sq. km.): 18,933 (49,036)
Population: 5,336,450
Ethnicity (%): Slovak (86); Magyar (10); Gypsy, Czech, other (4)
Languages: Slovak; Magyar; other
Religions (%): Roman Catholic (60); atheist (10); Protestant (8); other (22)
Type of Government: Parliamentary democracy

SLOVENIA
Area in sq. mi. (sq. km.): 7,834 (20,296)
Population: 1,967,700
Ethnicity (%): Slovene (91); Croat (3); Serb (2); other (4)
Languages: Slovenian; Serbo-Croatian; other
Religions (%): Roman Catholic (96); Muslim (1); other (3)
Type of Government: Parliamentary democracy

YUGOSLAVIA
(FEDERATION OF SERBIA AND MONTENEGRO)
Area in sq. mi. (sq. km.): 39,507 (102,350)
Population: 10,700,000
Ethnicity (%): Serb (63); Albanian (14); Montenegrin (6); Magyar (4); other (13)
Languages: Serbian; Albanian
Religions (%): Orthodox (65); Muslim (19); Roman Catholic (4); Protestant (1); other (11)
Type of Government: Federated republic

Map 2: Eastern Europe — Physical

Today we can define the geographical eastern border of Europe as the line formed by the combined western borders of Belarus, Ukraine, and the Russian Federated Republic. (See Map 1.) This is true because no concrete natural border exists to separate the Eurasian land mass into the two distinct geographical continents of Europe and Asia. Although apparently arbitrary, such a border can be justified by historical and cultural arguments.

The easternmost area of Europe, therefore, is characterized by two well-delineated geographical regions. The better defined of the two is the Balkan Peninsula in the south. It is bounded on the west by the Adriatic Sea, on the east by the Black Sea, on the southeast by the Aegean Sea, and its southern tip juts into the Mediterranean Sea. The Carpathian Mountains are the main land border to the north, although rivers, most notably the Danube, cut numerous gaps in this barrier. Close to 70 percent of the land in the Balkans is comprised of mountains. Except for predominantly narrow coastal plains, most of the lowlands in the Balkans' interior are deep river valleys that provide the peninsula with both a modicum of shallow layered arable land in the interior and its only lines of natural overland communication. Climatically, the peninsula is not a unit. It enjoys a Mediterranean-type climate along its sea coasts and a continental one throughout most of its interior. Thus in general, the Balkans can be characterized as having a harsh, divisive environment, which has played a decisive role in shaping the lives of those peoples who have inhabited the region since earliest recorded times. Historically, life in the Balkans has been far from stable and prosperous, and this has resulted in centuries of fierce competition among the peninsula's various populations for control of the geographically restricted available natural resources.

Immediately north of the Balkan Peninsula lies the other geographically definable region of Eastern Europe: the Danubian Basin. This region centers on the Danube River. The basin is essentially a broad, fertile plain, stretching from the point the Danube enters the Balkan Peninsula northwestward to Vienna. It is surrounded by mountain chains and plateaus on its peripheries. Within the folds of the chains lie two extensive mountain plateaus that serve as the basin's boundaries. In the southeast is the Transylvanian Plateau; in the northwest lies the Bohemian Plateau. Both are linked by river systems to the basin's central plain. The plain enjoys a continental climate and rich, fertile soil that have helped make it a thriving agricultural region for centuries. The surrounding mountains and hills are rich in ores, minerals, and timber, all of which have contributed to the growth of industry there. An extensive system of rivers draining into the Danube from every corner of the region has traditionally provided it with an exceptional communication and transportation network, which has, in turn, historically fostered political and economic unity in the area as a whole. The Danubian Basin served as the core territory of the Habsburg Empire from the 16th to 20th centuries and was one of the primary reasons for that state's continuous status as a European Great Power during that period. Following World War I, the dismemberment of the Habsburg Empire into small, independent nation-states demonstrated the importance of the Danubian Basin as a geographical entity. Carving up its territory among separate states completely disrupted the organic economic network of the basin that had thrived because of Habsburg consolidation. None of the new countries created in the basin was blessed with sufficient natural resources to guarantee a healthy economic, and thus political, existence.

Immediately north of the Danubian Basin the terrain falls off into a low, rolling plain extending some four hundred miles northward to the southern shores of the Baltic Sea. This strip of land is one section of the Great Eurasian Plain. Although it is bounded on the south by mountains and on the north by sea, no notable land features serve to define either its eastern or western borders. Two extensive river systems, the Oder in the west and the Vistula in the east, drain this Polish Plain. The geographical prominence of the open plain has given this third area of Eastern Europe its lasting name: Poland (from the Slavic word for field or plain, *pole* — thus, "Land of the Plain"). Its topography is characterized by a gradual transition from the mountains in the south to the plainland proper in the north. The southern uplands create a number of definable subregions, consisting of highland plateaus and surrounding lowlands, which possess nearly all the area's available fertile agricultural and rich mineral resources. These historically have been at the center of intense rivalries among the societies inhabiting the region. The undulating central plain and the flat northern coastal plain offer little by way of natural resources. The northern third of the plain is covered by thick pine forests that conspire against soil fertility. The northern continental climate is harsh, and the plain suffers from an utter lack of natural frontiers to both east and west. The strains of attempting to retain control of what amounts to almost unlimited open space has made independent political life for the societies inhabiting the plain precarious.

The three most identifiable geographical components of Eastern Europe are Southeastern Europe (the Balkan Peninsula), Central-Eastern Europe (the Danubian Basin), and Northeastern Europe (the Polish Plain).

BALTIC
SEA

Elbe R.

Niemen R.

Vistula R.

Oder R.

Bug R.

Pripet R.

POLISH
PLAIN

Dnieper R.

Ore Mts.

Sudet Mts.

*BOHEMIAN
PLATEAU*

Bohemian Forest

Šumava
Highlands

Morava R.

Tatra Mts.

Danube R.

Caroathian Mts.

Dniester R.

Julian Alps

DANUBIAN
BASIN

Bihor Mts.

Prut R.

Drava R.

Tisza R.

*TRANSYLVANIAN
PLATEAU*

Sava R.

Dinaric Alps

BALKAN

PENINSULA

Danube R.

BLACK
SEA

ADRIATIC SEA

Albanian
Alps

Morava R.

Balkan Mts.

Rila Mts.

Maritsa R.

Vardar R.

Rhodope Mts.

Pindos Mts.

*AEGEAN
SEA*

MILES
0   50   100 150 200

0   100   200   300
KILOMETERS

Taigetos
Mts.

*MEDITERRANEAN
SEA*

☐ Land from 1,500 ft (458 m) to 3,000 ft (915 m)   ---- Border of Eastern Europe's major regions
☐ Land from 3,000 ft (915 m) to 6,000 ft (1,830 m)
☐ Lands above 6,000 ft (1,830 m)

Map 3: Eastern Europe — Demographic

The demographics of Eastern Europe are diverse. While there are scattered small groups such as Gypsies, Vlahs, Jews, Italians, and Friulians, the East European population is primarily composed of four major groupings: Slavs, Germans, Turks, and indigenous peoples.

The Slavic peoples' ancestors entered Eastern Europe between the 5th and 7th centuries from their original common homeland, located somewhere north of the great Pripet Marshes. (See map 2.) Over time, the westward-moving Slavs developed distinct tribal cultures in their new habitats, leading to the emergence of separate Polish, Czech, Moravian, Slovak, and Sorb peoples. Collectively, these peoples are known as the West Slavs. Other Slavic tribal groups moved south and southwest from their Pripet homeland. These tribes roamed throughout Southeastern Europe, extending their migrations as far south as the Peloponnesian Peninsula of mainland Greece. The Slavs who settled in Southeastern Europe—Bulgarians (eventually), Croats, Montenegrins, Macedonians, Serbs, and Slovenes—came to be called the South Slavs. There was a third group of Slavic tribes that tended to remain near the original Pripet homeland; those who did migrate did so in an easterly or southeasterly direction, either into the forested regions of the northern Eurasian Plain or onto the vast, open steppes. The Slavs of this group—Great Russians, White Russians, Ukrainians, and Ruthenians—came to be known as the East Slavs.

Germans constitute a significant population in Eastern Europe. Heavy concentrations of Germans, representing the Austrian (or Southern) subgroup of the Germanic linguistic family, have existed in the western and northwestern corners of Central-Eastern Europe since early medieval times. Others have been native inhabitants of the fluid western border of Northeastern Europe. Starting in the 12th century, various groups of Germans pushed eastward into Eastern Europe, usually at the behest of non-German rulers, who desired to use them as military forces or as commercial colonists to stimulate trade and ore mining. The Prussians of Northeastern Europe and the Hungarian and Transylvanian Saxons of Central-Eastern Europe settled in this way. Likewise, other German colonies were established during medieval times in various areas of Serbia and Bulgaria in Southeastern Europe. Thus, by the close of the European Middle Ages, pockets of Germans could be found throughout Eastern Europe, from the shores of the Baltic Sea to the heart of the Balkan Peninsula. Although most of the Germans in the Balkans were eventually assimilated by their non-German neighbors, they have been, and continue to be, important demographic components in the remaining areas of Eastern Europe.

The Turkish presence has been significant in Central-Eastern and Southeastern Europe. In the former area, the Magyars (Hungarians) established themselves on the Pannonian Plain in the late 9th century. Unlike their nomadic Turkic predecessors, they did not disintegrate or disappear from the European scene, but instead consolidated their hold on the Pannonian Plain and established a strong, centralized state. Their permanent settlement in the heart of Eastern Europe had a decisive impact on the ethnic character of the region. They served as a dividing wedge that isolated the three Slavic linguistic subgroups from each other, causing the emergence of definitive linguistic differences among them. In Southeastern Europe, the Turkic Bulgars established a state south of the Danube in the late 7th century. By the 9th century, the Bulgars were becoming ethnically assimilated by their large Slavic-speaking subject population. Their conversion to Orthodox Christianity at mid-century resulted in their rapid and total assimilation. Within a century, most traces of their Turkic origins had disappeared, except for their name: the Bulgars had been transformed into Slavic Bulgarians. Other Turks, such as Uzes, Pechenegs, and Cumans, appeared in Southeastern Europe between the 9th and 11th centuries. Most of them eventually suffered the ethnic fate of the Bulgars, leaving only a few small, scattered ethnic traces of their presence along the coastal regions of the Black Sea. Furthermore, the Ottoman Turks' five-century rule over the Balkan Peninsula established scattered enclaves of Turkish-speaking groups throughout much of the area, with a heavy concentration in the region of Thrace.

Indigenous peoples have enjoyed a continuous existence in Eastern Europe. For example, the Southeast European Greeks precariously survived the Slavic invasion and settlement of their mainland Balkan possessions in the 6th and 7th centuries, and, by the late 10th century, won back much of the territory that was originally lost through Byzantine military efforts. Under the Ottomans, who granted the Greeks extensive commercial, political, and cultural privileges, Greek ethnicity acquired a solid foothold in sectors of Macedonia, Thrace, and Albania. Other surviving indigenous peoples in Southeastern Europe include the Albanians, who inhabit the central Adriatic coastal area of the Balkans and speak a unique language thought to be descended from ancient Illyrian, which would place them among the oldest groups of peoples in Europe (along with the Basques of northern Spain and southwestern France). Romanians, who occupy the northeast corner of the Balkans, claim a similar heritage. They speak a Latin-based language that, in Romanian national thinking, supposedly derives from over a century-and-a-half of Roman occupation in the ancient region of Dacia during the 2nd and 3rd centuries. They claim the ancient Dak people as their pre-Roman ancestors.

Map 4: Eastern Europe — Cultural

To make sense of Eastern European history, one must understand the cultural forces that have historically operated in that region. Human culture operates within societies on two levels, the macro and the micro. Macroculture, or civilization, is a complex culture shared by a network of numerous peoples spread over a large geographic area, in which all important social activities are encompassed by highly developed, sophisticated institutions; urbanization is widespread; a written language has been developed; and both division of labor and social differentiation exist. On the other hand, ethnonationality is now the primary expression of microculture for individual human societies. Every civilization incorporates a number of societies, with their respective microcultures, and the basic unifying factor is either a universal religious belief or a universal philosophy (or both in combination) to which all its membership subscribes.

Since the beginning of recorded time, interactions among human societies and their cultures have shaped and driven history. Every civilization has been forged and tempered by human contacts on the two levels of culture: Interrelationships with other civilizations (macrocultural contacts), and interrelationships among its own member societies (microcultural contacts). Of the human interplays that take place on each level, those involving civilizations possess the potential for the greatest degree of human upheaval.

In Eastern Europe today we are witnessing the interactions of three civilizations: the Western European (Catholic/Protestant), the Eastern European (Orthodox), and the Islamic (Muslim). To simplify our understanding of them, we might borrow the geological analogy of continental plate tectonics. Every civilization can be said to hold sway over a large, geographically defined core area in which the fundamental worldview binding together its constituent societies has undergone native, organic development. Each core area might thus be viewed as a large plate superimposed on the map of the world. Focusing on Eastern Europe, we find the Orthodox East European plate sandwiched between the Western European (to its west and northwest) and the Islamic (to its south and southeast). Their points of contact form cultural fault lines just as tectonic plates in similar fashion form seismographic fault lines.

But human fault lines cannot be delineated neatly on a map. Centuries of human interaction have occurred, and each microcultural society has penetrated into geographic regions dominated by others. Human cultural faults, therefore, can be seen as bands of green on the map whose widths vary by location and history. And much like geological faults, these bands represent lines along which occur the most dramatic disturbances, caused by friction among the differing macrocultural plates. Likewise, they are the points at which future eruptions are most likely to occur.

Historically, the cultural fault line dividing the West and East European civilizations runs from the Baltic to the Adriatic seas and roughly encompasses the three Baltic republics; the regions of Belarus and Ukraine bordering on Poland, Slovakia, and Hungary; Transylvania in Romania; the Vojvodina province of Serbia; the Slavonian border region separating Croatia and Serbia; all of Bosnia-Hercegovina; and the Dalmatian-Montenegrin border. A second fault line separates the Eastern European and the Islamic civilizations in the Balkans. Although seemingly short in the Balkans—this fault line parallels the border of Turkey with Bulgaria and Greece—it traces the extent of centuries of Islamic Ottoman rule and creates an extensive green band that resembles a long, somewhat scythe-shaped swath cutting northwestward into the Balkans through Bulgaria, Greece, Macedonia, Albania, and Kosovo, eventually intersecting the East-West European fault in Bosnia-Hercegovina. Therefore, the faults of the three civilizations of Eastern Europe all converge in Bosnia-Hercegovina.

Human frictions along and within the faults of Eastern Europe have been long-lived, with frequent flare-ups. Since the 9th century, the Baltic states, Transylvania, Banat, Vojvodina, Slavonia, Bosnia-Hercegovina, and the eastern borders of Poland have frequently witnessed human-cultural seismic eruptions. The East European–Islamic fault was formed in the 14th century and was pushed steadily northward by the Ottoman Turks until the 17th, after which time the fault was pressed to its present location in the early 20th century. Every region of the Balkans has served as a cultural flashpoint at least once, with Bosnia-Hercegovina, Macedonia, Kosovo, and Bulgaria proving the most recent.

Within Eastern Europe, the microcultural groups that are members of West European civilization are: a small number of Albanians, Croats, Czechs, Hungarians, Poles, Slovaks, and Slovenes. Member societies of East European civilization are: some Albanians, Bulgarians, Greeks, Macedonians, Romanians, and Serbs, as well as some Gypsies and Vlahs. East European Islamic societies are: most Albanians, Bosniaks, Turks, Pomaks (in Bulgaria, Greece, Macedonia, and Serbia), and some Gypsies and Vlahs.

LITHUANIA
TO RUSSIA

POLAND

BELARUS

GERMANY

West European

UKRAINE

CZECH REP.

East European

SLOVAKIA

AUSTRIA

HUNGARY

MOLDOVA

SLOVENIA

CROATIA

BOSNIA

ROMANIA

SERBIA

YUGOSLAVIA

BULGARIA

MONTENEGRO

ITALY

ALBANIA

MACEDONIA

East European

Islamic

GREECE

TURKEY

MILES
0   50   100   150   200

0   100   200   300
KILOMETERS

West European—East European fault

East European—Islamic fault

Area of convergence of all three civilizations

Part I

Early Medieval Period

(to the 13th Century)

Map 5: The Division of the Roman Empire, Late 3rd Century

By the late 3rd century, the Roman Empire was in its advanced stages of internal decline. The army, its numbers swelled with mercenaries hired to meet mounting German and Persian foreign threats, was all-powerful, making and unmaking emperors with bewildering rapidity and resulting in near-constant political anarchy. Moral decay infected the general Roman population, reflecting a widening social chasm between wealthy and poor that could not be countered by the proliferation of eastern mystery religions, whose numerous deities became trite and whose messages offered little hope for the future. The general population had scant interest in suffering the physical hardships entailed by active military service. The office of emperor was reduced primarily to that of supreme military commander, whose personal presence was required in the field to ensure the troops' loyalty and guarantee any chance of successfully defending the empire against foreign enemies. Under the circumstances, Rome had ceased being the empire's capital. Instead, the capital became wherever the emperor established his headquarters—Milan, Lyon, or Trier, among other locations. This situation made successful defense of the empire's borders against multiple threats highly problematic.

Emperor Diocletian (284-305), a common soldier by background, had the common sense of a professional military man. If success depended on the emperor's presence, then multiple threats required multiple emperors. He decided to divide the empire into two halves, western and eastern, each possessing its own emperor *(augustus)*. Each emperor would enjoy the services of a junior partner *(cæsar),* who also would be recognized as the heir-apparent. It was hoped that this would ensure both that supreme military command would be exerted adequately in the halves and that anarchy surrounding successions would be forestalled. Diocletian's administrative system, known as the *Tetrarchy,* worked only as long as he reigned. When he abdicated in 305, it quickly broke down, with the four "emperors" *(tetrarchs)* launching multisided civil wars (311-24) among themselves for political supremacy.

The line drawn by Diocletian dividing the Roman Empire into two administrative halves was neither arbitrary nor purely administrative. The culture of the empire was not strictly Roman, but rather a longstanding partnership with that of classical Greece. A dual quality permeated all aspects of Greco-Roman civilization. That duality was sustained through the use of both Greek and Latin in the Roman Empire, although the speakers of one considered those of the other culturally inferior. In general, Latin speakers inhabited the Roman provinces in the western portions of Europe and Africa, while Greek speakers predominated in the European Balkan, eastern North African, and Middle Eastern provinces. When Diocletian drew the primary west-east administrative division, he did so along the line that marked the invisible human cultural divide separating the Greek East from the Latin West in the northwestern corner of the Balkans. As a solution to the grave problems facing the empire at the end of the 3rd century, Diocletian's splitting of the Roman state ultimately proved a failure, but it resoundingly succeeded in institutionalizing the demarcation—creating the hyphen—between the two distinct branches of Greco-Roman civilization. After Diocletian, these branches developed along two increasingly divergent lines. Western European civilization has its roots planted in the western, Latin branch of Hellenism through the empire of Charlemagne, while Eastern European civilization is grounded in the eastern, Greek component by way of the Byzantine Empire.

The eastern state emerged during the reign of Emperor Constantine I the Great (306-37), ultimate victor in the civil wars among the *tetrarchs.* Constantine was convinced that his success was aided by the god of the Christians—a religious sect declared illegal by Diocletian but to which many of his troops secretly belonged. Unlike the Hellenic and the eastern mystery religions, Christianity possessed a strong ethical-moral system and the necessary sanctions for enforcing its precepts. It also provided its adherents with a consistent explanation for human life, grounded in the idea that all existence is a progression from the creation of the universe to ultimate spiritual salvation in life everlasting. Everything, including the Roman Empire, had its place in that continuum. Constantine recognized the utility of tying Christianity to the interests of a Roman state under his personal control. A state partnership with Christianity could help strengthen and perpetuate the empire by reinvigorating the morale of the population, crystallizing their loyalty, and unifying the Roman Mediterranean world. By Constantine's time, Christianity had begun to spread among the empire's German enemies, so that it was wielding influence beyond Rome's borders and might be used to expand Roman authority.

Constantine took an active part in establishing Christianity as a state religion. (Christianity later became *the* state religion under a successor, Emperor Theodosius I the Great [379-95].) For Christianity to serve Constantine's purposes, it needed to be unified in its beliefs, but the nature of the Christian deity—the Trinity—made that difficult. At the time, the church suffered from the Arian heresy, which revolved around differing concepts of the nature and substance of Christ (second entity in the Trinity). As Pontifex Maximus (one of the titles accompanying the imperial office), Constantine convened and directly participated in the First Christian Ecumenical Council of Nicæa (324) to settle the Arian issue. The result was the desired partnership of church and state and the emergence of the emperor as the sole authority figure in a now divinely ordained Christian Roman Empire.

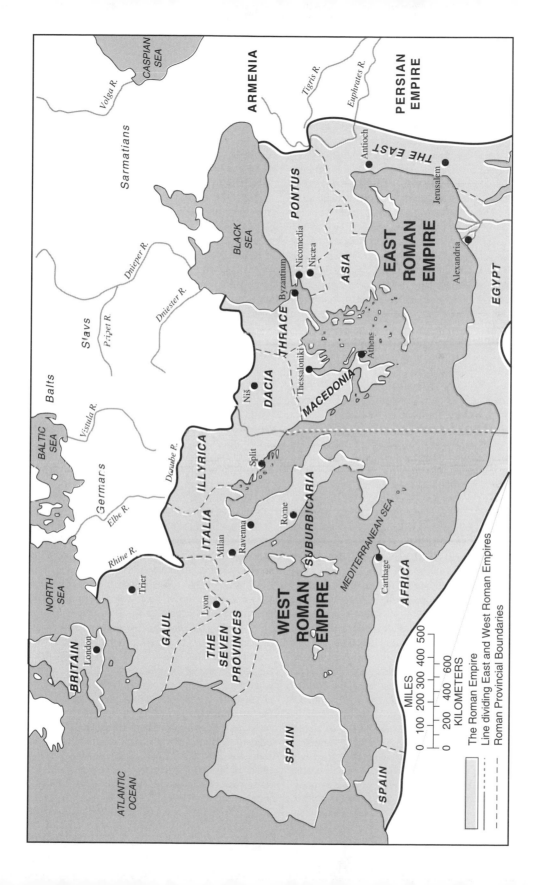

Map 6: The Barbarian Migrations, 4th–6th Centuries

In cultural-historical terms, "Europe" is the product of a human alloy composed of three elements: Greco-Roman traditions, Christianity, and "new" peoples (the so-called "barbarians" and their cultures). The first two components were present in the Roman Mediterranean world by the opening of the 4th century. The last was added by migrations of non-Roman societies into the territories of the Roman Empire during the 4th through 9th centuries. Without these, one can barely imagine Europe as anything but a geographical term.

Peoples of mostly Germanic origin inundated the western, Latin-speaking regions of the Greco-Roman world during the 4th through 6th centuries; they were joined in the 5th by Hunno-Turkic peoples immortalized in the person of their greatest leader, Attila. Attila and his Huns, though terrifying in their presence, exited the historical scene after only a brief appearance, but the Germans stayed. The German tribes were the westernmost in a long chain of peoples that stretched across the Eurasian steppes to Mongolia. When disturbances in China in the 4th and 5th centuries caused successive domino-effect waves to ripple westward along that human chain, the Germans lying just outside the northern borders of Rome had nowhere else to move except into the empire itself. At first, they merely attacked the eastern half, but eventually they overwhelmed the western. There they settled, decisively changing the demographics of the region.

The key factors in the German takeover were the chronic economic, military, and administrative weaknesses of the Western Roman Empire relative to the eastern half. Diocletian had recognized this situation at the time he divided his state — he chose the eastern portion for himself. Under pressure from the Huns, who were pushing westward on the Eurasian steppes, Germanic Visigoths (west Goths) successfully sought refuge within the Danubian borders of the Eastern Empire in 376. Dealing with an unprecedented situation, Emperor Valens (364-78) botched the administrative job of peacefully integrating the Visigoths into the empire. A war resulted in which the mounted Goths decisively crushed the Roman infantry and killed Valens in the Battle of Adrianople (378). Learning their lesson, successive eastern emperors conducted assimilatory policies (monetary and administrative bribery, interethnic marriages) to pacify the Visigoth leaders. In 401 Visigothic king Alaric (395-410) was named supreme Roman military commander and led his forces into Italy to face Stilicho (a Vandal contender for his military office who commanded most western imperial forces) and the growing numbers of other German tribes (including Burgundians, Vandals, and Ostrogoths) who were pushing across the borders of the empire, which had been stripped of troops by Stilicho and his supporters. Alaric proved victorious. Extorting power and wealth from the now militarily dependent western emperors, he sacked Rome in 410. His successors led the Visigoths out of Italy into southern Gaul and Spain, where a Visigothic kingdom was established in 419. Meanwhile, the Vandals had already settled in Spain, and under Gaiseric they pushed into Northern Africa, eventually capturing Carthage (439) and consolidating a kingdom that controlled the primary African breadbasket for Rome and the western Mediterranean. In 455 the Vandals then attacked and systematically sacked Rome itself.

The Ostrogoths entered the empire as Roman military allies in the 450s, settling first on the Pannonian Plain. Their king, Theodoric the Great (471-526), followed in the footsteps of Alaric by ravaging the Eastern Empire until he was bribed with the office of supreme military commander. In 489 he was then ordered to expel the Herulians from Italy. Their ruler, Odovacar, had abolished the office of Western Roman emperor in 476. After defeating Odovacar, Theodoric established an Ostrogothic state that paid lip service to Eastern Roman imperial authority but in practice was an independent kingdom. Theodoric issued coins and law codes in the name of the eastern emperor but cemented independent alliances with virtually all of the newly emerging Germanic kingdoms in the West.

The Franks, like most other Germanic invaders, entered the territories of the empire as its military ally in the late 4th century, settling south of the Rhine River in northern Gaul. Unlike the other Germanic intruders, they did not migrate as a people but, rather, expanded as a state. Frankish forces played a crucial role in defeating the Huns at the Battle of Châlons (451) and continued to fight as Rome's ally against Germanic Visigoths and Saxons. Under King Clovis (481-511) the Franks were converted to Roman Catholicism (the only Germanic people to be so; all of the others remained Arian heretics) and a powerful state controlling most of Gaul was consolidated. Before his death, Clovis was named an honorary Roman consul by the eastern emperor Anastasius I (491-518), which technically brought the Franks into the Roman Empire, but, in the West by that time, their subordination to East Rome was a fiction.

Most of the Germanic interlopers established states of their own, loosely modeled after the Christian Roman Empire. They retained a bastardized form of Latin Hellenism through the papacy and the Roman Catholic church, which had survived the disruption of the invasions and which served as the cultural cement that lent them cohesiveness. A number of the new states, especially the Ostrogoths', maintained the fiction of political subordination to the Eastern Roman emperor. All of them adhered to the Christian political culture forged at the Nicæan Council, accepting the notion of a divinely ordained church and state partnership, with a vague notion that the distant eastern emperor was God's supreme secular representative on earth.

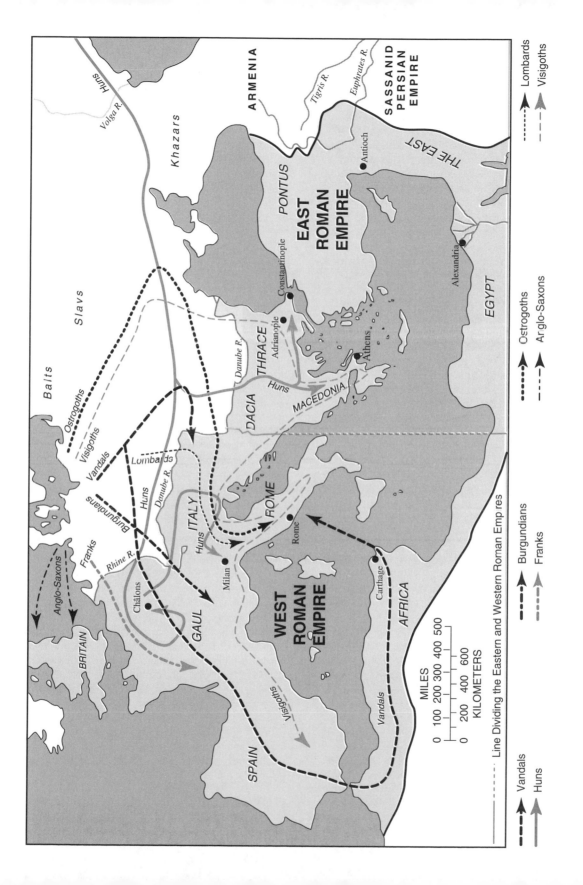

ARMENIA

SASSANID
PERSIAN
EMPIRE

Tigris R.

Euphrates R.

Volga R.

Huns

Khazars

PONTUS

EAST
ROMAN
EMPIRE

• Antioch

THE EAST

Balts

Slavs

Ostrogoths

Visigoths

Vandals

Lombards

Huns

Danube R.

Danube R.

THRACE

• Constantinople

Adrianople •

DACIA

Huns

MACEDONIA

Athens •

• Alexandria

EGYPT

Franks

Burgundians

Anglo-Saxons

Huns

Rhine R.

ITALY

Huns

ROME

Chalons •

Milan •

GAUL

BRITAIN

WEST
ROMAN
EMPIRE

Rome •

Carthage •

AFRICA

Vandals

SPAIN

Visigoths

MILES
0 100 200 300 400 500

KILOMETERS
0   200   400   600

- - - - - Line Dividing the Eastern and Western Roman Empires

Lombards

Visigoths

Ostrogoths

Anglo-Saxons

Burgundians

Franks

Vandals

Huns

Map 7: The East Roman Empire under Justinian I, Mid-6th Century

Following the Germanic migrations, the last concerted effort to reestablish a unified Christian Roman Empire within its classical borders was made in the mid-6th century by East Roman emperor Justinian I the Great (527-65). He came to the throne a somber and bookish man of humble origins, and seemed an easy target for manipulation by the popular factions in Constantinople. In 532 the factions attempted to force Justinian to make governmental changes and to reform taxes by rioting in the streets, setting fire to much of the city, and proclaiming his overthrow. (See Map 11.) Bolstered by his empress, Theodora, Justinian ruthlessly ordered his military to crush the uprising. Over the bodies of 30,000 of his subjects, Justinian emerged as a ruler determined to solidify the autocratic power of his office and to realize the grandiose pretensions implied by the official imperial ideology of the post-Nicæan Christian Roman Empire: one God, one Emperor, one Empire.

For the rest of his reign, Justinian relentlessly pursued his imperial goals in both domestic and foreign policy to the point of driving his empire into near bankruptcy by the time of his death. He ordered the great systemizing of Roman law that has since borne his name, and that served as the fundamental legal basis of state government in much of both Eastern and Western Europe until the late 18th century. He sponsored expensive building campaigns, with much effort and many resources spent on both practical civic projects and a new Christian church architecture to express the power and glory of the state-church partnership. Belief in the mystical Christian imperial ruler was reinforced through the borrowing of numerous cultural elements from the neighboring Persian Empire. These served to ritualize autocratic political authority and to fashion a conclusive Christian façade onto the Hellenic tradition of deified Roman emperors.

With this powerful, mystical Christian aura, Justinian began asserting his claim to supreme earthly authority over the Christian world-state. When the western church, acting in the name of the Nicæan Christian community of Italy, appealed to Justinian for relief from Arian Ostrogothic rule, he responded with a massive military effort to wrest North Africa—the primary source of agricultural food products for the West—from the Arian Vandals, and Italy from control by the German Ostrogothic heretics. Justinian set out to politically recreate the unified Roman Empire, with its new capital centered at Constantinople and himself as the undisputed Christian autocratic monarch.

From 533 until 554 Justinian waged near-constant warfare in the West. In 533 the capable general Belisarius was despatched with a small military force to North Africa. He swiftly crushed the Vandals, led by Gelimer, and the imperial forces gained control of nearly the entire region during the following decade, eventually pushing their control as far as southern Spain. Meanwhile, in 535 Belisarius led an invasion of Sicily, overran it, and moved into southern Italy, which he soon conquered from the Ostrogoths. Rome was in imperial hands by late 536. Ostrogothic king Witiges then subjected the city to a year-long, but unsuccessful, siege, after which Belisarius advanced into northern Italy, took Ravenna (the Ostrogoths' capital) and captured Witiges. Unfortunately, he was then recalled to Constantinople, and the new Ostrogoth ruler, Totila, rolled the imperial forces back into southern Italy by 543. Three years later, Totila captured and sacked Rome, prompting Belisarius to return and retake it, only to abandon it again in 549 to the Ostrogoths. Justinian replaced Belisarius as field commander with Narses, who invaded Italy from the north with a large army of mostly Hunnic and other barbarian mercenaries. He destroyed the Ostrogoths in 552 and brought all of Italy under Justinian's control.

By the conclusion of the Italian wars (554), Justinian's empire had reincorporated Italy, Sicily, most of North Africa, and the southern coastal regions of Spain in the West. The Vandals and the Ostrogoths were utterly defeated and consigned to history's dustbin. Administrative and religious unity of East and West was reestablished, and, over it all, Justinian briefly reigned supreme.

Justinian's reconquests in the West proved to be the swan song for a unified Christian Hellenic world-state. The financial and human costs of imperial success had been prohibitive. Both the imperial treasury and the manpower of the military were exhausted. Two decades of conflict had left Italy ravaged. Its non-German inhabitants, at first thankful for their liberation from heretical German rule, swiftly grew to resent the centralized, efficient taxation system imposed on them by the imperial government in Constantinople. Their economic woes were magnified by their growing awareness of the cultural differences that separated them from their East Roman liberators. While both were Nicæan Christians, the Italians rapidly sensed that Greek- and Latin-speakers, though outwardly adhering to the same beliefs, did not possess identical perceptions of their common faith. The practical, nononsense Latins soon came to resent the presence of the Greeks in Italy, whom they saw as being heartless, cunning, and altogether unscrupulous. Fourteen years after Justinian's victory, in 568, a new Germanic tribal people, the Lombards, invaded Italy. The Latins offered the depleted and weak imperial defense forces little assistance in resisting the onslaught, and within four years the empire had lost most of the Italian peninsula, except for Rome, Ravenna, and the southern regions.

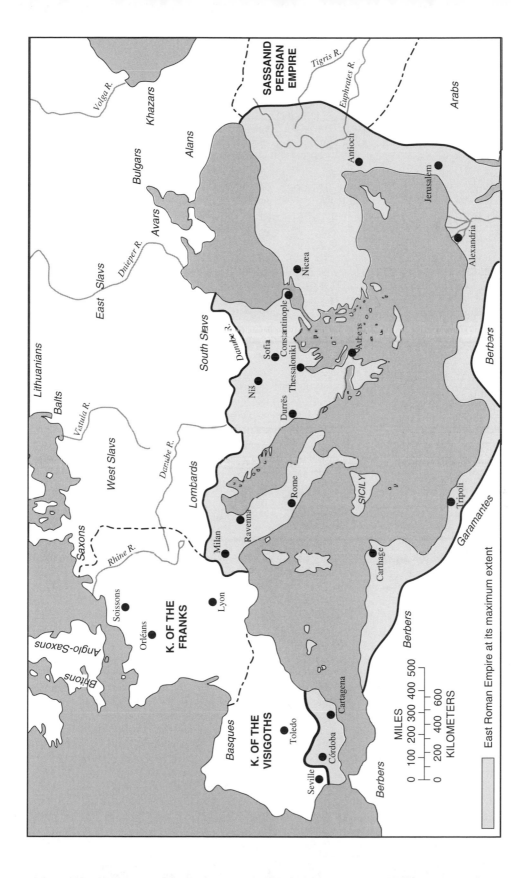

SASSANID PERSIAN EMPIRE

Arabs

Khazars

Volga R.

Bulgars

Alans

Tigris R.

Euphrates R.

Antioch

Jerusalem

Alexandria

East Slavs

Dnieper R.

Avars

Nicæa

Lithuanians

Balts

Vistula R.

South Slavs

Danube R.

Sofia

Constantinople

Niš

Thessaloniki

Athens

Durrës

Berbers

West Slavs

Danube R.

Lombards

Rome

SICILY

Saxons

Rhine R.

Milan

Ravenna

Tripoli

Garamantes

Carthage

Briton

Anglo-Saxons

Soissons

Orléans

K. OF THE FRANKS

Lyon

Berbers

Basques

K. OF THE VISIGOTHS

Toledo

Cartagena

Córdoba

Seville

Berbers

MILES

0   100 200 300 400 500

0   200   400   600
KILOMETERS

East Roman Empire at its maximum extent

Map 8: Slavic and Turkic Invasions, 6th–8th Centuries

Emperor Justinian died in 565 before the futility of his western policies was fully revealed, but signs of insuperable strain in maintaining them had arisen even prior to his death. Already Avar and Slavic forces had penetrated the Balkan possessions of the Eastern Empire, in a prelude to the inundations that followed over the next 150 years.

The Avars, a Turkic people from Central Asia, entered Europe in the 550s, pushed westward until they were stopped by the Franks, and then occupied the lowlands of the Danubian Basin, which they consolidated as their home power base for raids against the Kingdom of the Franks and the East Roman (Byzantine) territories in the Balkan Peninsula. Their attacks on the latter region, commencing in the 550s, were particularly savage and no area of the peninsula was spared. Numerous Avar parties not only raided but began to settle south of the Danube, causing the empire increasing military and administrative turmoil. Thessaloniki, the second-largest commercial port in the empire, was repeatedly besieged by the invaders, and the plains outside Constantinople were periodically laid to waste. Imperial control virtually ceased to exist in the western Balkans. In 629, during the reign of Byzantine emperor Heraclius (610-41), the Avars, in league with the Persians, besieged Constantinople but suffered a grave defeat from which they did not adequately recover. Various non-Avars who had been brought under Avar sway successfully rebelled, and in 678, following the defeat of the Arabs before Constantinople, the surviving Avars requested Byzantine suzerainty. Charlemagne destroyed the remains of the Avar Danubian state between 795 and 796.

The most numerous and loyal allies of the Avars, from the beginning of their raids into the Balkans until the defeat before Constantinople, were various Slavic tribes. The Slavs' original homeland is thought to have been the region drained by the Pripet River on the fringe of the Eurasian steppes. From that place numerous Slavic tribal groups expanded—a response to the demographic pressures of the barbarian migrations—in search of new homelands to the west, south, and east. Many of the southern-moving Slavic tribes linked themselves—either by force or by choice—as infantry allies to the mounted Avar forces that dominated the Danubian Basin and the northern stretches of the Balkan Peninsula. They then accompanied them on their raiding and colonizing incursions into the Balkan lands of the Byzantine Empire. While scattered Avar groups established settlements within the borders of Byzantium, their Slavic allies, in search of new homelands, did so as a matter of policy, decisively transforming the ethnic demography of the eastern and central areas of the region. The indigenous populations were forced either to flee to the coastal lands of the peninsula, where they stood a chance for imperial protection, or to remain in the interior, where they existed for a time in isolated urban or rural pockets before suffering eventual ethnic extinction through assimilation with the Slavs.

Emperor Heraclius did what he could to protect the refugees and to win back control of the peninsula's interior, but he was handicapped by a major war with the Persian Empire. Though able to recover much of the former Greek lands from the Slavs, he was forced to invite into the Balkans Croat and Serb Slavic forces to serve as border guards in the region's northwest. In return, he was compelled to grant them settlement rights and they swiftly filled the imperial vacuum in the Balkan west, thus augmenting the numbers of Slavic inhabitants. Direct Byzantine control in the western, central, and northern Balkans would never again be completely secure.

Most Slavic tribes that moved west, north of the Danubian Basin, also found themselves placed under Avar control in areas vacated by the westward-migrating Germans. When Avar power declined after their defeat before Constantinople, the Czech, Moravian, Slovak, and assorted other Slavic tribes settled on the northwestern periphery of the Danubian Basin and were briefly united in the mid-7th century into a loose confederation by a certain Samo, who may have been a Frankish tradesman. New evidence seems to demonstrate that "Samo's state," long considered a political union by some historians (and by most Slovak nationalists), may have been only a more militant-than-normal trading union under Samo's overall control. He was able to expel the Avars from the region and to repulse Frankish attempts by King Dagobert (628-38) to gain control of his operations. When Samo died in 658, the tribal union under his authority—whatever that may have been—disintegrated.

In 679 another Turkic steppe people crossed the mouth of the Danube and established a base of operations on its southern banks. The Bulgars, under their ruler Asparuh (d. 701), defeated the Byzantine border forces and won a treaty with the empire in 681 that recognized a Bulgar state in former Byzantine territories south of the Danube. The flatlands north of the river were also under Bulgar control and joined to the lands ceded by the Byzantines. The resulting Bulgar state represented the first officially recognized "barbarian" state in Eastern Europe and it quickly became Byzantium's principal rival for political hegemony in the Balkans.

The Turkic Bulgars successfully exerted their political and military authority over the Slavic tribes that already inhabited the former Byzantine lands granted to them by the treaty. Unlike the Avars, who apparently maintained a segregation between themselves and subservient non-Avar subjects, the Bulgars integrated the Slavic tribal leadership into their ranks (in a secondary position), opening the door to a certain level of ethnic intermingling that would eventually lead to their ethnic assimilation by the Slavic majority.

Balts

Polochanians Krivichians

Germans
Obodrites
Veletians
Pomeranians

Oder R.
Polanians
Vistula R.

Dregovicians

Radimichians

Mazovians
Bug R.

Pripet R.

Dnieper R.

Severians

K. of
FRANKS

Sorbs
Elbe R.
Silesians

Czechs

SAMO'S
STATE

Vistulans

White Croats

Derevlianians

Volhynians

Polianians
Ulichians

Danube R.
Augsburg

Moravians Slovaks

Dniester R.

Tivertsians

Prut R.

Carinthian
Slavs

Tisza R.

AVARS

LOMBARD
KINGDOM

Pannonian
Slavs

Drava R.
Sava R.

Olt R.

BULGARS

EXARCHATE
OF RAVENNA

Ravenna

Croats

Split

Serbs

Danube R.

Pliska

Rome

LOMBARD
KINGDOM

Naples

Morava R.

Sofia

Maritsa R.

B
Y
Z
A
N
T
I
N
E

Vardar R.

Plovdiv

Constantinople

Thessaloniki

EXARCHATE OF RAVENNA

E
M
P
I
R
E

Athens

MILES
0   50 100 150 200

0   100   200   300
KILOMETERS

The Bulgar State, 701
The Avar State, 7th Century
Samo's State, mid-7th Century
Border of the Byzantine Empire
Lands of Baltic Peoples

Map 9: Eastern Europe, Mid–9th Century

Two large independent states had emerged in geographical Eastern Europe by the middle of the 9th century—Bulgaria (also known as the First Bulgarian Empire) and Great Moravia (or the Great Moravian Empire). Both were under increasing political and cultural pressures from two powerful states seeking to expand their authority in the region—the Byzantine Empire (or East Roman Empire) and the Kingdom of the Franks (or the Holy Roman Empire). The earliest independent East European state was Bulgaria.

Following Bulgaria's establishment in the late 7th century, successive rulers increased both its size and its significance in the foreign affairs of the Byzantine Empire (see Map 10). By the reign of Prince Boris I (852-89), Bulgaria had consolidated such a hold over large regions of the northern and central Balkans that it posed a decisive challenge to Byzantium for political hegemony in the peninsula. But Bulgaria faced potential internal problems caused by continued ethnic disunity between a Turkic Bulgar ruling establishment that was declining in number and a majority subject Slavic population that adhered to varying paganistic religions and was increasing in number with every expansion of the state's borders. Moreover, combined with the near-constant threat of Byzantine military might, Frankish efforts to gain a foothold along the Adriatic coast of the Balkans increased pressure on Bulgaria in the distant northwest. Croat and Serb tribal leaders in that area countered the Franks through loose and unstable alliances with Bulgaria, Byzantium, or the Franks, but these obviously were undependable for guaranteeing stability in that sector.

Into this situation rose the Great Moravian state. Since the end of Samo's union, a power vacuum had existed in the region. There, a Czech-Moravian tribal leader, Mojmir (833-36), succeeded in shaping a new Slavic tribal union centered on the Morava River basin. Rostislav (846-69) expanded the new state until it was viewed by both the Franks and the Bulgars as a potential threat. Moravia experienced its apex under Prince Svatopluk (870-94), who extended control northward and southward into Pannonia.

The Franks exerted great pressure on Moravia, especially through missionary activity that won increasing successes within the general population. Realizing that Christianity at the hands of the Franks would mean ultimate German ecclesiastical and political control, Rostislav petitioned Byzantium for religious and political aid. Emperor Michael III (842-67) responded in 863 by dispatching a missionary effort of his own to convert Moravia to Orthodox Christianity, thus removing it from German Catholic control.

The mission was headed by two brothers, Cyril and Methodius. Prior to his departure for Moravia, Cyril, a linguist, philosopher, and diplomat, devised a written alphabet (Glagolitic) for Slavic, based on the Slavic dialect spoken near Thessaloniki and formed by a mixture of Greek, Phoenician, and assorted other eastern letters. Once in Moravia, the two missionaries proved so successful that the German Catholic church leadership vehemently denounced them to the pope. (At that time, the split between Catholic western and Orthodox eastern Christianity had not yet attained official institutional recognition—that would come with the so-called Great Schism of 1054.) The row instigated by the Germans became known as the "Three Language Heresy." The German church claimed that only Greek and Latin were acceptable "sacred" languages with divine authority to convey the Christian Scriptures. In Rome, Cyril successfully defended his new Slavic alphabet at the papal court, stressing that, while Greek and Latin were both pagan in origin, Glagolitic was Christian from its inception. Cyril died in Rome before having a chance to return to Moravia. Methodius was promoted to bishop of Moravia and did return, but proved unable to resist the mounting German pressure on his diocese. Upon his death (885), Methodius's followers were expelled from Moravia by German church authorities, ensuring German cultural penetration of the state.

Meanwhile, in 865 Boris in Bulgaria decided to convert his state to Christianity to gain recognition for Bulgaria as a member of the Christian European community of states and to stave off possible extermination by either Byzantium or the Franks for being an alien pagan threat. Furthermore, conversion would unify the divided population. When the Byzantine Empire threatened Bulgaria's borders with an army and a fleet, demanding that Boris convert to Orthodoxy, Boris did so, after which point Greek clergy entered Bulgaria. Conversion immediately subordinated Bulgaria to Byzantine cultural dominance.

Boris's son and successor, Simeon I (893-927), realized that religious-cultural subordination to Byzantium would eventually lead to political subservience. When Cyril and Methodius's Moravian disciples appeared on Bulgaria's borders after being expelled from their homeland, Simeon quickly capitalized on this windfall. He lavished rich patronage on them and supported their efforts to provide his state with Slavic liturgical works and the training necessary to build a native Bulgarian church hierarchy. From Simeon's farsighted policy, and the work of the original disciples and their Bulgarian protégés, sprang a definitive Slavic literary culture based on a new, simplified alphabetical system—Cyrillic. Mostly composed of modified Greek characters and named in honor of Cyril, the language was ingeniously precise in representing all of the phonics found in the Slavic languages. The original Greek liturgical texts in Bulgaria were translated into Slavic using the new letters, and native Bulgarian clergy were trained in Cyrillic. The Greek hierarchy in Bulgaria was replaced by a native Bulgarian organization, providing Simeon with the cultural component to ensure Bulgarian independence.

Map 10: The Rise of Bulgaria, 8th–10th Centuries

The treaty of 681 with Byzantium granted the Bulgars territories south of the Danube and north of the Balkan Mountains. From their capital at Pliska, the Bulgar *hans* (rulers) controlled the Slavic inhabitants of their newly-acquired lands, as well as Wallachia and other lands north of the Danube stretching northeastward to the Eurasian steppes. Under the successive heirs of Asparuh, the new state's founder, Bulgaria (also known as the First Bulgarian Empire) attempted to expand its territories in the Balkans at Byzantine expense, either through peaceful and favorable alliances with the imperial authorities (such as Tervel's [701-18] with Emperor Justinian II [705-11]) or through warfare. By the opening of the 9th century, Bulgar ruler Krum (808-14), who from his capital at Pliska ruled a large state stretching to Great Moravia in the north, was in the position to commence life-threatening attacks on the Byzantine Empire.

After crushing Avar forces in the Danubian Basin (803), Krum was determined to establish absolute autocracy when he emerged as Bulgar ruler in 808. Throughout his reign, Krum conducted all-out warfare against Byzantium for control of the Balkans. Although his military successes at first fluctuated, his final years generally were successful. In 811 the Bulgars massacred a Byzantine army led by Emperor Nicephorus I (802-11), who was killed and whose head was ordered fashioned into a silver-lined ceremonial drinking cup by Krum. During the next two years he swept the Byzantines from northern stretches of the Black Sea coast, decisively defeated another imperial army, and appeared before the land walls of Constantinople, briefly causing a panic in the imperial capital. Krum bequeathed to Bulgaria expanded territories in the central and eastern Balkans, including the Byzantine city of Sofia.

Krum's immediate successors consolidated Bulgaria's gains through an extended peace treaty with Byzantium before initiating conquests in the central and western Balkans. Prince Boris I continued to push his borders westward but met sharp resistance from Germans in the Croatian northwest (853) and from Serbs (860), many of whom then came under direct Bulgar control. Boris also established Ohrid as an important Bulgar administrative center in Macedonia, and he expanded his control over western Thrace in the south until an outlet was opened on the Aegean Sea. But Boris is most significant for having initiated the conversion of his mixed Turkic and Slavic subjects to Orthodox Christianity. The conversion resulted in the creation of a literary form of the Slavic language that has had crucial historical and cultural consequences to the present day. (See Map 9.)

The strength of the large, autocratically centralized Bulgar state, which represented the collective legacy of Krum and Boris, was fully realized by Boris's successor, Simeon I. Under his reign, Bulgaria achieved its historical apex, encompassing virtually all of the Balkan Peninsula except Croatia, Thessaloniki, Greece, and the Thracian environs of Constantinople. His royal patronage of the literary activity by Moravian refugees and their Bulgarian students in the capital at Preslav, as well as his personal literary interests (Simeon had been educated while held hostage by the Byzantines in Constantinople) resulted in the Cyrillic alphabet's creation, the single most important Slavic cultural development in medieval Europe. His reign is termed Bulgaria's "golden age." (See Map 9.)

Like Krum, Simeon conducted near-constant warfare with Byzantium. Within a year of his succession (894), he was able to place a cohesive and experienced military force in the field that wrested control of most of the Balkans from the empire. Bulgarian military power forced Byzantium to recognize the Bulgarian church's autonomy (a Bulgarian patriarchate was established in Ohrid), though technically it was still considered by the Byzantines to be subordinate to the Greek patriarch of Constantinople. The Bulgarian ruler was recognized as second in power and authority only to the emperor in Constantinople. Four times between 913 and 924 Simeon advanced to the land walls of the imperial capital but was unable to breach them. Simeon proclaimed himself emperor—*tsar*, in Slavic—of the Romans and Bulgarians (924), and his claim was taken seriously. Ultimate success appeared just within his grasp, and only the impregnable strength of Constantinople's defenses prevented Simeon from being acknowledged as emperor of Byzantium.

While occupied to the south with operations against the Byzantines, Simeon was compelled to deal with military problems on other fronts. Starting in 917, Byzantium attempted to militarily weaken Bulgaria by inducing Balkan Slavic and steppe nomadic Turkic peoples to attack Bulgaria from the north, thus opening operations against Simeon on two fronts simultaneously. The Magyars, a Turkic people, swept out of the steppes and permanently wrested all Bulgarian territories north of the Danube, as far as Pannonia, from Simeon's authority. Serbs, urged on by the Byzantines, rebelled against the Bulgarians in 918, and in 924 Simeon conquered and ravaged Serb lands formerly outside the Bulgarian state. Despite their efforts, the Byzantines never managed to successfully contain Bulgarian expansion in the central and southern Balkans while Simeon was alive. He died in 927 an imperially frustrated but powerful regional ruler.

Bulgaria's leadership in the Slavic world, earned by Simeon's military and cultural endeavors, was short-lived. It fell swiftly into political decline following his death. By the close of the 10th century Bulgaria was reeling from Russian and Byzantine invasion, and in 1018 it was finally crushed by the Byzantines and wholly incorporated into their empire for the next 167 years. (See Map 15.)

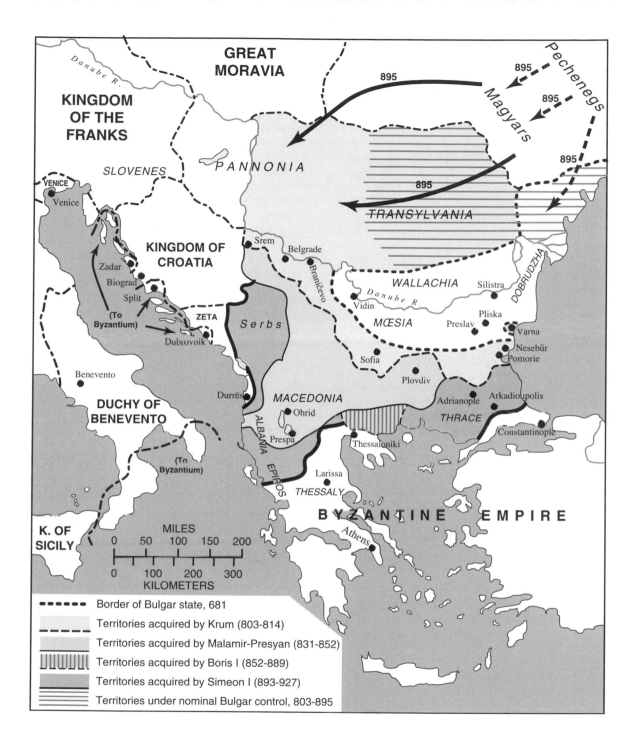

GREAT MORAVIA

KINGDOM OF THE FRANKS

Danube R.

895

Pechenegs

895

Magyars

895

895

SLOVENES

PANNONIA

VENICE

Venice

KINGDOM OF CROATIA

Zadar

Biograd

Split

(To Byzantium)

ZETA

Dubrovnik

Benevento

DUCHY OF BENEVENTO

(To Byzantium)

K. OF SICILY

Serbs

Durrës

MACEDONIA

Ohrid

Prespa

ALBANIA

EPIROS

Larissa

THESSALY

Athens

Srem

Belgrade

Braničevo

TRANSYLVANIA

WALLACHIA

Silistra

DOBRUDZHA

Vidin

Danube R.

MŒSIA

Preslav

Pliska

Varna

Nesebŭr

Pomorie

Sofia

Plovdiv

Adrianople

Arkadioupolis

THRACE

Constantinople

Thessaloniki

BYZANTINE EMPIRE

MILES
0   50   100   150   200

0   100   200   300
KILOMETERS

● ● ● ● ● ●   Border of Bulgar state, 681

– – – –   Territories acquired by Krum (803-814)

Territories acquired by Malamir-Presyan (831-852)

Territories acquired by Boris I (852-889)

Territories acquired by Simeon I (893-927)

Territories under nominal Bulgar control, 803-895

Map 11: Constantinople, 10th–12th Centuries

For the God-ordained Christian Roman Empire forged at the Council of Nicæa, Emperor Constantine I dedicated a new capital city in 330. Erected on the site of the ancient Greek colonial port of Byzantium, New Rome, or Constantinople (Konstantinopolis, or City of Constantine, in Greek) was located on an easily defended triangular bit of land on the European shore of the Bosphorus Strait. Directly to its north was the crescent-shaped mouth of a small river emptying into the Bosphorus, known as the Golden Horn, which formed the only natural harbor in the area. To its south stretched the Sea of Marmara, which emptied into the Mediterranean through the Dardanelles Strait. The Bosphorus-Marmara-Dardanelles seaway separated Europe from the West Asian landmass and linked the interior of Eurasia beyond the Black Sea directly to the Mediterranean. Byzantium, with its sea walls and the Golden Horn, dominated traffic on this crucial seaway. Moreover, the city was located on the most direct overland pathway between the two continents. Through fortifications of this land triangle, Constantinople was transformed into an impregnable fortress-city that controlled all the important lines of communication between the European and Asian worlds. Until the 13th century, it was the largest, strongest, wealthiest, and most culturally developed city in all of Europe.

Constantinople symbolized the Orthodox Christian Roman imperial order itself. Within its walls, though civic structures often intentionally mimicked their counterparts in Old Rome, there was no place whatsoever given over to paganism. New Rome, with its massive triple land walls and encircling sea walls, was considered the God-protected capital for all of His chosen Christian people, home to the supreme imperial Christian emperor, and, in the guise of the patriarchal cathedral of Hagia Sophia, heart of the Eastern Orthodox church. The city was a material symbol of the Nicæan imperial partnership of state and church, a partnership that was believed destined to last until the end of the world. (In fact, the land walls, constructed in 411 by Emperor Theodosius II [408-50], for over a thousand years were never breached by a foreign enemy, and the sea walls, with a few notable exceptions over the centuries, withstood naval threats mounted by Persians, Arabs, and Russians.)

From its founding, Constantinople grew and was continuously embellished by successive Byzantine emperors. By the 10th and early 11th centuries, under the patronage of the Macedonian imperial dynasty—who brought Byzantine culture and political-military power to its zenith—the capital was filled with splendid civic structures, including palace complexes, the Hippodrome arena, public forums, aqueducts, water cisterns, churches, and monasteries. The highly de-

veloped municipal government and the Orthodox church ministered to the social and health needs of the inhabitants, who were a mixed bag of Greeks, Armenians, Bulgars, Jews, Turks, Russians, Khazars, Egyptians, Italians, and others drawn from all corners of the empire. Residents benefited from organized fire brigades; plumbing and sewage systems; free hospitals; and other amenities virtually unknown in the rest of Europe at the time, such as schools, street lighting, and public baths. Sophisticated trading regulations, extensive docking facilities on the Golden Horn, public forums, and a system of open-air markets all nurtured commerce. Numerous monasteries provided food, shelter, and health care to the homeless and destitute.

Public life in Constantinople was centered around the Augustæum forum, where the two primary imperial palace complexes (the Great and the Mangana), the patriarch's magnificent basilica of Hagia Sophia, and the huge circus arena of the Hippodrome were located. The palace districts were replete with impressive buildings and parks, and the Mangana housed an excellent university that emphasized studying classical Hellenism when it was virtually unknown in Western Europe, as well as advanced Orthodox theology. Hagia Sophia, considered the supreme symbol of Orthodox Christianity, provided the model for religious art and architecture throughout the empire, and its numerous trained staff constituted a pool for filling positions both in the upper church hierarchy and in missions intended to spread Byzantine culture to peoples beyond Byzantium's borders (e.g., the Bulgars, Serbs, and Kievan Russians). The popular craving for spectacle was satisfied by frequent events in the Hippodrome (including chariot races, hunts, mystery plays, and acrobatics) staged by the imperial government. These became the focus of public participation in Byzantine political affairs; for example, the widely popular sports factions, whose memberships were identified by colors, evolved into loose political parties that expressed themselves through mass demonstrations and, in extreme cases, street riots. (See Map 7.)

Constantinople, popularly considered the "queen of cities," was both the epitome of Byzantium and an enormous drain on imperial resources. It defined all official political and cultural trends but its dominant position could be maintained only by consuming a large proportion of the empire's fiscal and human assets. This resulted in a progressive decline within the empire's provinces and a growing rift between the provincial aristocracy and the government functionaries in the capital. By the first quarter of the 11th century, strains between the capital and the provinces led to a series of civil wars that weakened the empire militarily and ended the Macedonian dynasty.

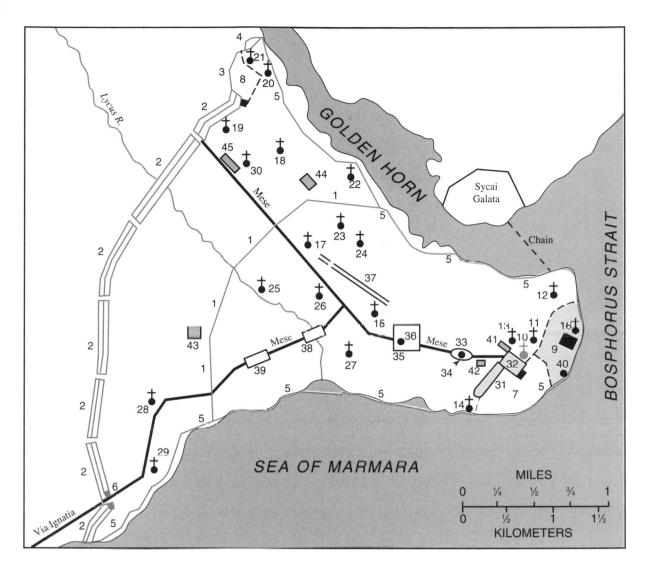

Walls

1. Constantine I (early 4th century)
2. Theodosius II (early 5th century)
3. Manuel Comnenus (mid-12th century)
4. Heracleus (7th century) and Leo (9th Century)
5. Sea walls
6. Golden Gate

Palaces

7. Great Palace District
8. Blachernæ Palace District
9. Mangana Palace District

Churches and Monasteries

10. Hagia Sophia
11. St. Irene
12. St. Demetrius
13. St. Mary Chalcoprateia

14. SS. Sergius and Bacchus
15. St. Savior Akataleptus
16. St. George of the Mangana
17. Holy Apostles
18. St. Mary Pammakaristos
19. St. Savior in Chora
20. SS. Peter and Mark
21. St. Mary of Blachernæ
22. St. Theodosia
23. St. Savior Pantepoptes
24. St. Savior Pantocrator
25. Constantine Lips
26. St. Polyeuctus
27. Myrelaion
28. St. Andrew
29. St. John of Studius
30. St. Prodromus in Petra

Civic Structures

31. Hippodrome
32. Augustæum
33. Forum of Constantine
34. Column of Constantine
35. Forum Tauri
36. Arch of Theodosius I
37. Aqueduct of Valens
38. Forum Bovi
39. Forum of Arcadius
40. Lighthouse
41. Basilica Cistern
42. Cistern of Philaxenus
43. Cistern of St. Mocius
44. Cistern of Aspar
45. Cistern of Ætius

Map 12: The Balkans, Early 11th Century

On Simeon's death in 927, the Bulgarian throne passed to his son Petŭr I (927-69). Petŭr was a pious Orthodox Christian and well-intentioned ruler, but personally he was weak. His piety helped end the incessant warfare with Byzantium instigated by his father. He cemented his friendly relations with the empire by marrying the granddaughter of Emperor Romanus I Lecapenus (920-44), who reciprocated by officially recognizing both Petŭr's imperial title and the autocephaly of the Bulgarian Orthodox patriarchate of Ohrid. But peace with Byzantium did little to strengthen a Bulgarian state led by such a pacific ruler as Petŭr. Throughout his disinterested reign, Bulgaria's powerful military forces were permitted to decay, leading to constant threats by the Magyars, who had established their own state north of Bulgaria's northwestern Danubian frontier, and to incursions by raiders from the Eurasian steppes in the northeast.

Under Petŭr, a widening chasm between the wealthy and powerful Bulgarian military-ecclesiastical aristocracy and the over-burdened and poor subject population resulted in growing social unrest. This was expressed through popular religious discontent, which was displayed by a sharp increase in monasticism (embodied in the national saint, Ivan Rilski) and the rise of the Bogomil heresy—a radically dualistic faith probably inspired by heretical Paulicians settled near Plovdiv (whom the Byzantines earlier had transferred from their native homeland in Armenia)—that emphasized total rejection of all worldly institutions and social order.

Early in the reign of Tsar Boris II (969-72), Bulgarian military and social decline led to the state's invasion and conquest by Kievan Russians, led by their prince, Svyatoslav (964-72). Boris was captured and made a puppet ruler by the Russians in 969. Threatened by the Russian presence, Byzantine emperor John I Tzimiskes (969-76) attacked the Russians (and their forced Bulgarian allies) in 972 by both land and sea. After ravaging the old Bulgarian capitals and pinning Svyatoslav and his troops against the Danube, the Byzantines forced the Russians to evacuate Bulgaria. Tzimiskes compelled Boris to abdicate, reduced the Bulgarian patriarchate of Ohrid to an archbishopric under the authority of the Greek patriarchate of Constantinople, and incorporated eastern Bulgaria into the Byzantine Empire.

Although Tzimiskes officially abolished the Bulgarian state, the defeated state's aristocracy continued to control the western Bulgarian lands, from the Danube in the north to Macedonia in the south, which had been relatively unaffected by the Russian invasion and subsequent Byzantine conquest. In 976 one of the Macedonian warrior-aristocrats, Samuil (976-1014), installed himself as ruler over the western Bulgarian regions. After establishing his primary capital in Ohrid, Samuil expanded his authority to the area around Sofia and proclaimed the reestablishment of the Bulgarian Orthodox patriarchate of Ohrid. His military forces raided deep into the Greek regions of Byzantium to the south, and through further efforts he won temporary control over territories extending to the Black Sea in the east.

Emperor Basil II (963-1025), the last powerful Macedonian Byzantine ruler, earned his sobriquet *Boulgaroktonos* (the Bulgar-Killer) in a series of crushing military campaigns (996-1014) conducted against Samuil. Unable to face the Byzantine forces in open battle, the Bulgarians resorted to guerilla-type warfare. Their tactics proved successful enough to trap and defeat Basil's army in the mountains southeast of Sofia in 981 and to stave off inevitable defeat for almost two decades. But Basil's relentless reduction of Bulgarian strongholds, and his effective policy of bribing numerous Bulgarian commanders to defect, gradually undermined Samuil's continued guerilla resistance. With Basil eating away at his territory, and with growing defections among his own commanders, in 1014 Samuil was forced to make a stand against the Byzantines in the eastern Macedonian mountains. His forces were outflanked and decisively defeated. As a gruesome demonstration of his victory, Basil ordered some 14,000 Bulgarian prisoners blinded, except for one out of every hundred, who was spared an eye to lead the rest back to Samuil's headquarters. The sight of the pathetic remnants of his army was said to have killed Samuil. Within four years, what was left of Bulgaria submitted to Byzantine rule and was incorporated directly into the empire. The Bulgarian aristocratic leadership was eventually acculturized by the Byzantines and the Bulgarian patriarchate reduced to an autonomous archbishopric under the Greek patriarch's authority.

To Bulgaria's north during this period, the Croat and Serb tribes founded respective states of their own. A certain Tomislav (923-28) united a number of Croat tribes under his authority and won recognition as king from the pope (who took every opportunity to increase papal influence in the Balkans at the expense of the Orthodox patriarch of Constantinople), who even permitted the use of Glagolitic among the Croat clergy (though Latin ultimately emerged as the dominant literary-liturgical language of the Croat church). Byzantium, which constantly sought strategically-placed allies against the Bulgarian threat, also recognized Tomislav. Thereafter, the new Croat state found itself in constant conflict with both Byzantium and Venice for control of lands along the Adriatic coast, and with Hungary over territories on the Danubian Basin's plain.

A seed for a small Serbian tribal state had been planted in the late 8th century but was stifled by Bulgarian and Byzantine incursions. With the disappearance of the Bulgarian state in the early 11th century, the way was opened for the Serbs to rebuild one of their own. This process was begun during mid-century in the small, mountainous Adriatic coastal region of Zeta.

Map 13: The Rise of Hungary, 10th–13th Centuries

The Magyars, a Turkic people speaking a Finno-Ugric language related only to Finnish and Estonian, were driven off the Eurasian steppes by the Pechenegs and entered Central-Eastern Europe in the late 9th century. Led by tribal chiefs of the Árpád family, they established themselves in seven tribal concentrations in Pannonia, which they used as a home base for raids deep into the rest of Europe, destroying Great Moravia in the process. The violence and frequency of their raiding reawakened the Europeans' memories of the terrors once inspired by Attila and his Huns, earning for the Magyars a new, and lasting, designation among their victims—Hungarians. Unlike their Turkic predecessors, however, the Magyars did not disintegrate or disappear when German imperial forces led by Holy Roman emperor Otto I the Great (936-73) decisively defeated them outside Augsburg in 955. Instead, under Duke Géza I (972-97), they consolidated their holdings in Pannonia, secured control of the mountain defenses surrounding it (including the Transylvanian Plateau and its Carpathian ramparts) and opened their borders to German Catholic missionaries. By the beginning of the 11th century the Magyars were converting to Catholicism, possessed a strong, centralized Central-East European state, and were recognized as members of the medieval European community—the pope even conferred a royal crown on their ruler, King St. István I (997-1038), in the year 1000.

István, a staunch Catholic and an exceptional ruler, suppressed the Orthodox missionary activity that had begun previous to his reign and crippled traditional Magyar paganism, to which many tribal chiefs still clung. He confiscated the lands of the broken tribal chieftains, along with various uninhabited lands, and transformed them into royal domain, distributing some to the Catholic church and organizing the rest as counties under centrally-appointed counts. Border defenses, especially along the Carpathians in the north and east, were systematically created and manned to provide István's state with both protection and springboards for state consolidation in Transylvania, Slovakia, and Ruthenia. István intentionally destroyed the Magyars' traditional Turkic tribal social system, replacing it with one modeled after the Frankish system of the Holy Roman Empire. Prior to his death, István fended off a German bid to win control of the Magyar state.

István's death sparked a reaction against his radical changes in Magyar society. He left no direct heir, so the Doge of Venice and István's nephew, Pietro Orseolo, briefly reigned (1038-46). But Orseolo's pro-German and pro-Italian policies got him expelled, and a period of unstable and transitory successions marked by German invasion and a final pagan uprising followed, until the Árpáds secured control of the throne starting with András I (1047-61). The new ruler threw back a series of German incursions, forcing the Holy Roman Empire to recognize Hungary's independence in 1058, and began to reestablish and consolidate István's system of royal centralization, a policy that was continued by his successors.

Hungary's internal order was restored under King St. László I (1077-95), who supported the papacy in the investiture conflicts with Holy Roman emperor Henry IV (1056-1106) and encouraged agricultural and commercial prosperity. In 1091 he conquered Croatia-Slavonia and Bosnia, in an attempt to gain direct access to ports on the Adriatic Sea, but was drawn into a see-saw struggle with Venice over control of the coastline and its hinterlands. László permitted his Balkan conquests considerable self-government under the loose oversight of a Hungarian-appointed *ban* (governor). His successor, Kálmán I (1095-1114), managed to wrest the coastline of Croatia from the Venetians in 1102, and the Croat-Slavonian nobility was constrained to accept a dynastic union with Hungary.

The so-called *Pacta Conventa* of 1102, the document of union between Hungary and Croatia-Slavonia (now believed to have been a forgery), became the focus of controversy. The Hungarians came to view the union as a conquest that led to Croatia's direct incorporation into the Hungarian state, while the Croats claimed it was a voluntary act that permitted the Croats to retain most of their political, legal, and social autonomy in exchange for recognizing the Hungarian monarch as ruler, thus creating a dual kingdom of Hungary-Croatia. Though the issue remains a bone of contention, there was no question that, after 1102 until the early 20th century, the fate of Croatia-Slavonia was linked to Hungary. Bosnia's relationship with Hungary remained far less direct.

Most of the 12th century was a period of dynastic instability for Hungary, yet during that time the state's defenses, economy, and control of conquered territories were strengthened by the colonization of Saxon Germans, Székelys, and other peoples, especially in Transylvania. But weakness on the throne resulted in the rise in the state administration of a powerful class of royal functionaries, who eventually evolved into a strong, numerous aristocracy. In 1222 they forced King András II (1205-35) to grant them the Golden Bull, which became a charter of feudal privilege and an important constitutional document for Hungary. It bestowed tax exemptions on functionaries and clergy, assured them an annual assembly (diet) for airing grievances, forbade foreigners (including Jews) from holding office, and granted landowners full rights to their property and freedom from arbitrary royal confiscations and arrest. The bull significantly limited royal authority and elevated the political influence of its recipients. Its negative impact was demonstrated during the reign of King Béla IV (1235-70), when a rift between the ruler and his functionary-landowners left a weakened state at the mercy of a Mongol invasion (1241).

POLAND

Prague

GREAT MORAVIA

Cracow

KIEVAN RUSSIA

Danube R.

Augsburg

HOLY ROMAN EMPIRE

SLOVAKIA   Košice

RUTHENIA

Bratislava   **X**   Nitra

Vienna   **X**

Esztergom   **X**

Székesfehérvár   **X**

Ljubljana

X

Venice

Zagreb

X

Kalocsa   PANNONIA

Tisza R.

Danube R.

X   TRANSYLVANIA

SLAVONIA

X   **X**

CROATIA

Nin
Belgrad

BOSNIA

Belgrade

Pechenegs

Danube R.

PAPAL STATES

Split

ZETA   SERBIA   Raška

Niš

BULGARIA

BENEVENTO   Dubrovnik

- - - - - - - - Border of Great Moravia, Late 9th Century

X   Area of Magyar Concentration, 10th Century

Hungary under István I, 1038

Territories added to Hungary by the mid-13th Century

MILES

0   50   100   150   200

0   100   200   300

KILOMETERS

Map 14: The Rise of Poland, Late 10th–13th Centuries

The Poles emerged abruptly into history in 963. In that year Saxon Germans, pushing eastward, stumbled onto an existing, well-organized Slavic state that until then had been unknown to anyone in the Germanic European world. The state was ruled by a tribal chieftain named Mieszko I (ca. 960-92), who bore the title of *piast* (a name originally meaning "second-in-command," but which Mieszko's family used so successfully to establish its rule over the various tribes within the state that it became the family's dynastic name). A shrewd individual, Mieszko quickly realized that the Poles' paganism made them vulnerable to increasing pressures from the Christian Germans, who could use the missionary cause as an excuse to win political domination and control of Polish territory. Mieszko pulled the rug out from under the Germans by forging contacts with the Roman papacy through the already Christianized Slavic Czechs of Bohemia. In 965-66 the Poles converted to Roman Catholicism and placed themselves directly under papal protection, thus assuring the Polish state's continued independence from the Germans.

The papacy welcomed the Poles' allegiance since it increasingly needed allies on the Holy Roman Empire's eastern flank against the German emperors for recognition of ultimate temporal authority in medieval Western Europe. To help ensure Polish independence from the Germans, and to further solidify their loyalty to Rome, the papacy around the year 1000 granted the Poles their own Catholic archbishop under Rome's direct authority, thus removing any German ecclesiastical-political control over Poland. The Polish archbishop's seat in the town of Gniezno also soon became the capital of the Piast rulers.

Mieszko's son and successor, Bolesław I the Brave (992-1025), brought stable administration to Poland. He attempted to create a large West Slavic state under his authority, briefly acquiring the Czech Bohemian throne (1003-4) and moving against Kievan Russia in the Ukraine. For the first time, Poland broke the land lock of the Great Polish Plain. At his death, a six-year dynastic struggle ensued, in which Czech Přemysls briefly occupied the Polish throne in their turn. Under the Piast Bolesław II the Bold (1058-79), Poland was immersed on the papal side in the investiture conflicts between Rome and the Holy Roman Empire, and, under Bolesław III Wrymouth (1102-38), the near-constant wars between Poland and the German Holy Roman Empire resulted in Polish independence from any claim of control by the German emperor. Wrymouth pushed the borders of his state northward to the shores of the Baltic Sea, Christianized the pagan Pomeranians, and incorporated their territories.

At Wrymouth's death, however, Poland was plunged into internal chaos over the royal succession. Part of the problem lay in the nature of the Polish royal crown itself. Unlike the hereditary crown granted the Magyars in 1000, the papacy bestowed on its Polish royal allies only personal crowns that recognized the authority of individual wearers alone. Each new ruler had to apply to Rome for renewed recognition of his authority. The uncertainty of royal legitimacy had repercussions in the matter of succession to the throne.

In Poland the system of succession to a personal crown had not been stabilized by Wrymouth's death. The Poles were still an essentially tribal people, though almost two centuries of full participation in the affairs of Catholic Western Europe had instigated political and social developments that were moving Polish society beyond tribalism. Western notions of primogenitor percolated among the Piasts and competed with the Poles' more traditional ideas of family seniority for determining the heir to the throne. The seniority approach prevailed in most tribal societies but proved highly disruptive because of the ensuing sibling rivalries that it caused.

The seniority principle held in Poland. Wrymouth, through his contacts with the Kievan Russian state in western Ukraine, attempted to standardize it by adapting the Kievan rotary system of succession. Wrymouth had five sons and, using the Kievan model, divided his Polish kingdom into five regions. He retained the town and region of Cracow as the royal capital and principality. When he died, it was expected (based on seniority) that his eldest son would inherit Cracow and, along with it, recognized leadership of the state. To reflect the new ranking, his other sons would each then rotate their holdings among the remaining four politically prioritized regions according to their ages.

Just as the rotary seniority system eventually broke down in Kiev, representing an innate weakness of an essentially tribal state vulnerable to more centralized foreign enemies (in the Kievan case, the Mongol-Tatars), its Polish manifestation failed almost immediately. After Wrymouth's death, instead of the rotary system's stabilizing the tradition of seniority, the state was plunged into anarchy. His sons had a difficult time accepting the political realities of the rotary regions—they saw their principalities as personal property rather than expressions of rank in the succession system—while their own sons set out to subdivide their fathers' regions further. The various family members' conflicts over the system wracked the Polish state with internal disunity. For the following 150 years, although certain Piast family members laid claim to the personal royal crown, it was difficult to tell who actually ruled in Poland. In the midst of the internal political discord, Poland was devastated by two Mongol invasions (1241 and 1259), which led to the invited colonization by Hanse Germans and the Teutonic Knights in the north to help rebuild, redevelop, and protect territories hard hit by the eastern intruders. (See Map 19.)

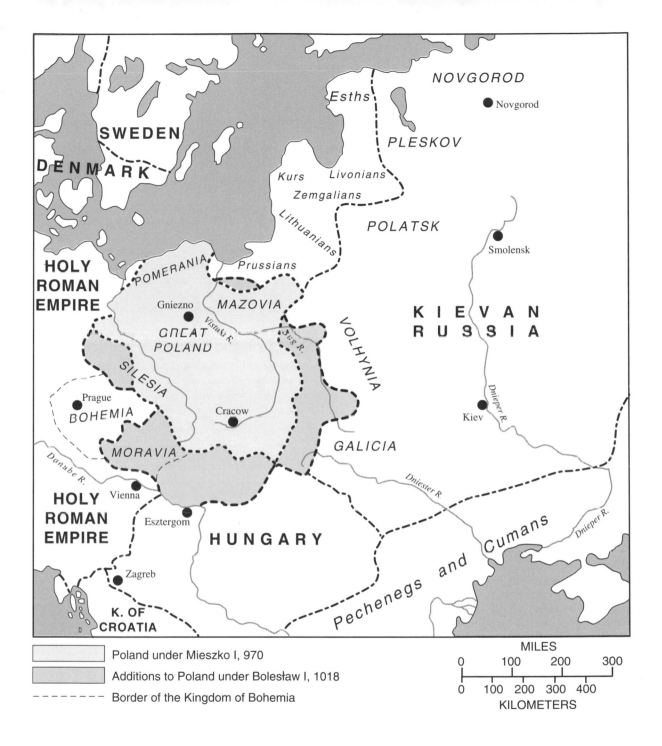

SWEDEN

DENMARK

HOLY
ROMAN
EMPIRE

NOVGOROD

● Novgorod

Esths

PLESKOV

Kurs

Livonians

Zemgalians

Lithuanians

POLATSK

● Smolensk

POMERANIA

Prussians

MAZOVIA

Gniezno
●

*GREAT
POLAND*

Vistula R.

SAN R.

VOLHYNIA

**K I E V A N
R U S S I A**

SILESIA

Prague
●

BOHEMIA

Cracow
●

MORAVIA

Dnieper R.

Kiev
●

GALICIA

Danube R.

Vienna
●

HOLY
ROMAN
EMPIRE

Esztergom
●

H U N G A R Y

Dniester R.

Dnieper R.

Zagreb
●

**K. OF
CROATIA**

Pechenegs and Cumans

MILES

| 0 | 100 | 200 | 300 |

KILOMETERS

| 0 | 100 | 200 | 300 | 400 |

▢ Poland under Mieszko I, 970
▨ Additions to Poland under Bolesław I, 1018
– – – – – Border of the Kingdom of Bohemia

Map 15: Eastern Europe, Mid–11th Century

For the first time in history, Eastern Europe in the mid-11th century was spanned by independent states stretching from the shores of the Baltic to those of the Mediterranean. While some of those states, such as Croatia and Zeta, would prove transitory in their existence, all of them would serve as the foundations for claims advanced by many East European nationalists in the 19th and 20th centuries.

In Poland, Bolesław I the Brave bequeathed to his successors territories stretching from the Baltic Sea and the western Carpathian Mountains. The kingdom was cemented together by an administrative system staffed by *castellans* holding both civil and military authority and by a structured Catholic church organization based on a network of Benedictine monasteries. Although his incursions into Bohemia-Moravia and Kievan Russia proved mostly temporary, and over a decade of succession problems and pagan anti-Catholic reaction followed his death, the Polish state ultimately managed to retain the internal unity imparted it by Bolesław. This was largely the work of Casimir I the Restorer (1038-58), who definitively reestablished Catholicism and returned order to the state's administration. But he succeeded at a price—the powerful and jealous Polish aristocracy forced him to renounce the royal title for that of grand duke and to grant numerous concessions to them and the clergy, establishing a precedent that would prove extremely damaging to the authority of Polish rulers.

Great Moravia was destroyed by the Magyars' arrival in the Danubian Basin. Thereafter, much of its former territories were prey to the rising Hungarian and Polish states, as well as to the militantly expansionary Holy Roman Empire. In Bohemia, the house of Přemysl consolidated a Czech successor state with their capital at Prague. Despite the efforts of Prince St. Václav I (d. 929)—the "jolly king Wenceslas" of Christmas carol fame—Bohemia was internally divided between Catholics and pagans, whose bitter conflicts, which were reflected within the ruling dynasty itself, rendered the state weak. Boleslav I (929-67) was forced to accept the suzerainty of the Germanic Holy Roman Empire, even though he brought Moravia under his control and consolidated royal authority over the Czech tribal leaders. His son, Boleslav II (967-99), achieved the final victory of Catholicism in Bohemia and actively supported missionary activities aimed at converting the Magyars and Poles, but continuing rivalries within the dynasty resulted in Polish king Bolesław I the Brave's brief occupation of the Bohemian throne following his death. The Poles were driven out by German forces and Bohemia remained in anarchy until the reign of Bretislav I the Restorer (1034-55), who expanded his control into Poland and was briefly Polish king. Alarmed by Bohemia's turnabout, Emperor Henry III the Black (1039-56) invaded and forced Bretislav to renounce his Polish gains and become a tributary of the Holy Roman Empire.

Like Bohemia, Hungary was in the throes of internal instability caused by the dynastic problems following the death of King István I (1038). (See Map 13.) Here too, conflicts between newly converted Catholics and those holding on to traditional paganism exacerbated the situation. The Árpád ruling family was factionalized among the throne's contenders, some of whom resorted to calling on German support for their cause. Emperor Henry III dabbled periodically in the Magyars' succession squabbles, at first supporting King Pietro Urseolo, who was also Doge of Venice, in return for an oath of fealty. In 1046, when Pietro was overthrown by Vatha (the Magyar tribal chief who led the last great pagan uprising in Hungary), and András I gained the throne and restored royal power, Henry led a series of three military campaigns against Hungary in an effort to exert his direct authority over that state. András held him off every time, forcing Henry in 1058 to recognize Hungary's independence from the empire.

Under the last powerful emperor of the Macedonian imperial dynasty, Basil II *Boulgaroktonos,* the Byzantine Empire enjoyed revived dominance over the Balkans by the mid-11th century. Bulgaria was crushed (1014) and completely integrated into the empire (1018), with its territories divided among the administrative-military *themes,* which were both recruitment areas and home bases for specific Byzantine military units, and which served as the empire's primary provincial organization. *Theme* commanders and their staffs wielded both administrative and military authority, and the former Bulgarian aristocracy was incorporated into their ranks. The efficiency of Byzantine administration, especially in taxation, led to a serious Bulgarian revolt in 1040, led by Petŭr Delyan, who proclaimed himself tsar and won mass support from the population in the central and north-central Balkans. But a power struggle with a certain Tihomir weakened his leadership and the Byzantines crushed the revolt in 1041. Soon thereafter, former northeastern Bulgarian regions of the empire were ravaged by Pecheneg raids (1048-54), while the general population became increasingly agitated by the spreading Bogomil heretical movement, which came to assume a certain anti-Byzantine sentiment.

In the Balkan northwest, both the Croats and the Serbs established states of their own. Croatia fell under Roman Catholic spiritual authority, briefly enjoying its own independent diocese. Its rulers, though able to expand their authority over Bosnia, were stymied in the north by the southward-moving Magyars and along the Adriatic coast by the Venetians. The Serbs had accepted Orthodoxy from their former Bulgarian overlords. By the mid-11th century, they had consolidated a small state in the highlands above the Adriatic coast known as Zeta.

Lithuanians

Smolensk

Obodrites

Veletians

Prussians

Gniezno

POLAND

KIEVAN RUSSIA

HOLY ROMAN EMPIRE

Prague

Cracow

Kiev

BOHEMIA

GALICIA

Danube R.

Augsburg

Slovaks

Ruthenians

Vienna

Esztergom

HUNGARY

Pechenegs

Zagreb

Danube R.

Venice

CROATIA

Belgrade

PAPAL STATES

Belgrad

BOSNIA

SERBIA

Danube R.

Pliska

D. OF SPOLETO

ZETA

Dubrovnik

B Y Z A N T I N E

BULGARIA

Rome

Normans

Sofia

Saracens

Bari

Ohrid

Plovdiv

Thessaloniki

Constantinople

E M P I R E

Saracens

Athens

MILES
0 50 100 150 200

0 100 200 300
KILOMETERS

Map 16: The Balkans, Late 12th Century

In 1054 the formerly unified European Christian church was torn into two halves—western Roman Catholic and eastern Orthodox—by the Great Schism. The break was the culmination of centuries of rivalry between both the Roman papacy and the patriarchate of Constantinople for spiritual supremacy within the church and the Byzantine and Holy Roman emperors for secular hegemony over a theoretical, and nonexistent, universal Christian world state. Although technically a religious matter, the schism sealed a cultural division between east and west expressed in mutual political animosity and ethnoreligious bigotry, the consequences of which have persisted into the present.

The powerful Macedonian dynasty of the Byzantine Empire had slipped into rapid decline after Basil II, with the provincial military aristocracy of the *themes* pitted against the bureaucratic functionaries of the capital for dominant influence over Basil's successively weak successors. Civil wars broke out in the mid-11th century, and the imperial throne alternated between champions of the provincial military, on the one hand, and functionary factions, on the other, until the able military commander Alexius I Komnenos (1081-1118) led a successful revolt, securing the throne and establishing a stable dynasty (1081-1185).

By the time Alexius returned stability to the imperial office, the empire had suffered frightful military blows at the hands of the Normans in Italy (the last Byzantine foothold was lost in 1071) and the Seljuk Turks in Anatolia (the Byzantine army was irreparably crushed at the Battle of Manzikert in 1071 and much of the region lost). Alexius initiated much-needed judicial and fiscal reforms, won the support of provincial military commanders, and counterbalanced their growing power by playing the high Orthodox clergy against them. He foiled the Italian Normans' attempted invasion of the Balkans (1085), suppressed the Bogomils' revolt in the former Bulgarian lands (1086-91), and blunted Cuman incursions south of the Danube (1091). Realizing his military inability to evict the Seljuks from Anatolia, Alexius requested western allies from Pope Urban II, thus launching the Crusades.

Alexius received far more than he bargained for in the Crusades. A succession of western crusading armies marched through the Balkans on their way to Constantinople, from which place they desired to sweep south through Anatolia and liberate the Holy Land from its Muslim Turkish masters. The undisciplined crusaders moved like a horde of human locusts through Serbia and the Bulgarian lands of the empire, cutting swaths of destruction and death as they traveled. Though Alexius managed to shuttle each succeeding crusader force into Asia with the least possible delay, he and his successors were unable to prevent or to repair the internal disruption caused by them in the empire's Balkan possessions. This situation, coupled with imperial preoccupation with the crusaders' activities in Asia, resulted in the reemergence of an independent Bulgarian state in 1185 and solidification of Serbia under a new royal dynasty by 1196.

In 1185 two brothers of possible Vlach origin, Petŭr and Ivan Asen, revolted against the empire. Though initially defeated and forced north of the Danube, they returned in 1186 with a Cuman army and compelled Emperor Isaac II Angelos (1185-95) to sign a truce giving them control of lands between the Danube and the Balkan Mountains. In 1189, the Asens raided deep into the empire, sparking Angelos to invade their lands in response. The Bulgarians soundly defeated Angelos and forced him to accept the independence of the resurrected Bulgarian state (also called the Second Bulgarian Empire). Both brothers were eventually murdered by members of their unruly aristocracy (Ivan in 1196, Petŭr in 1197), and their younger brother, Kaloyan (1197-1207), succeeded them. He signed a peace with Byzantium (1201) and launched successful campaigns against Serbia and Hungary, extending Bulgaria's western borders.

In Serbia at this time, Stefan Nemanja (ca. 1167-96), a *župan* (local ruler) from the Raška region, managed to unite various clans into a functioning state, which he held together through Orthodox Christianity. Bogomil heretics, who had entered his lands from Bulgaria, were persecuted and forced into Bosnia. Stefan proclaimed his complete independence from Byzantium and expanded his territories to the west and south, incorporating the older Serbian state of Zeta. He abdicated in 1196 and retired to Hilandar Monastery on Mount Athos, which his son, St. Sava, had founded. (See Map 18.)

Bosnia had long been a bone of contention between Croats and Serbs, with control passing back and forth between them, until an independent Bosnian state emerged in the late 12th century under the rule of a certain Kulin (ca. 1180-1204). He acknowledged nominal Hungarian suzerainty—while working to attain full independence by developing Bosnia's silver mining industry and establishing lucrative trade arrangements with Dubrovnik—and also parried the powerful cultural-political influences of Roman Catholicism and Orthodoxy by tolerating Bogomilism.

Catholic Hungarian and Orthodox Serbian rulers, eager to gain Bosnia for themselves, accused Kulin and his family of being devoted to the Bogomil belief. They asserted to the pope that Kulin was attempting to make Bogomilism the state religion of the Bosnian lands under his authority, lands which were considered Roman Catholic because of their past Hungarian-Croatian connection. But when Pope Innocent III preached a crusade against Kulin and the heretics, the Bosnian ruler wisely felt constrained to announce his adherence to Catholicism and to permit a Catholic synod in Bosnia condemning Bogomilism (1203). Despite Kulin's supposed recantation and the decision of the Catholic synod, the heresy continued to spread among the general Bosnian populace.

KIEVAN
RUSSIA

Danube R.

HOLY
ROMAN
EMPIRE

Vienna

Esztergom

HUNGARY

Kalocsa

TRANSYLVANIA

Cumans

Zagreb

SLAVONIA

Venice

To
Venice

CROATIA

Belgrade

Vlahs and Cumans

PAPAL
STATES

Zadar

Split

BOSNIA

SERBIA

Danube R.

Vidin

Raŝka

BULGARIA

Tŭrnovo

Varna
(To
Byzantium)

Dubrovnik

Duklja

Sofia

KINGDOM OF THE
TWO SICILIES

Naples

Durrës

MACEDONIA

Plovdiv

Adrianople

Ohrid

Serres

Constantinople

Thessaloniki

Ioannina

BYZANTINE   EMPIRE

SICILY

MILES

0   50   100 150 200

0   100   200   300

KILOMETERS

Athens

CRETE

	Border of Second Bulgarian Empire, 1197
	Border of resurrected Bulgarian state, 1187
	Territories under direct Bulgarian control
	Territories under loose Bulgarian control

Part II

Late Medieval Period

(13th –15th Centuries)

Map 17: The Latin Empire of Constantinople, 1214

Catastrophe struck the Byzantine Empire in 1204. Venetian mariners, commanded by Doge Enrico Dandolo (1193-1205), the central figure and mainspring of the undertaking, and French knights, led by Boniface de Montferrat, descended on Constantinople in 1203. They constituted most of the Fourth Crusade's military forces, originally despatched against the Muslim Turkish and Arab masters of Egypt and, then, the Holy Lands by Pope Innocent III. Normans from the Kingdom of the Two Sicilies augmented their ranks. From pope to common warrior, the crusade was riddled with enemies of the Orthodox Byzantine empire.

The cultural animosities engendered by the Great Schism, combined with fateful circumstances, caused the crusade to end with the capture and sack of Constantinople rather than the assault of Muslim Egypt or Palestine. The crusading army that gathered at Venice (1202) in response to Pope Innocent's call proved smaller than expected, but its leaders had contracted ships in advance based on their original estimates of need. Dandolo convinced Montferrat that the smaller forces could make up Venice's costs by helping him capture the Croat Adriatic port of Zadar. Once this was accomplished, Dandolo managed to divert the forces to Constantinople. He wanted Venice to win greater commercial advantage in the eastern Mediterranean over Byzantium, to which the city had long been subordinated. Alexius Angelos, son of recently deposed Emperor Isaac II Angelos (1185-95), aided Dandolo in his scheme by appearing outside Zadar bearing letters from a contender for the western imperial throne—Philip of Swabia, son-in-law of the deposed Isaac and brother of late Holy Roman emperor Henry VI (1190-97)—requesting the crusaders help restore Isaac to the Byzantine throne.

Under pretext of restoring Isaac, in return for his financial and other material support of their crusading efforts, the crusaders descended on Constantinople in 1203. Although successful in their short-term goal, their champion proved too much a creature of the Orthodox East to play the role of lackey for Catholic Westerners, whom the Byzantines looked upon (with some justification) as barbaric and less culturally developed. In 1204 the Western Crusaders turned their greed and frustration on the Byzantines, whom they considered (just as justifiably) to have reneged on promises of support for their endeavor. The Crusaders eventually broke through the sea walls protecting Constantinople and, once inside, gave free rein to a venomous cultural animosity that has been rivaled in Southeastern Europe only by the carnage displayed in the latest warfare in Bosnia. The wholesale rapine, pillaging, and plundering inflicted by the Catholic warriors of Christ upon the stricken capital of the East, the largest and wealthiest city in the world at the time, were unprecedented. The Orthodox world never forgot or forgave the West for the sack of Constantinople, and the event created a gulf between the Eastern and Western European civilizations that has existed virtually unbridged into the present.

The Latin Empire was able to plant few roots in the hostile Orthodox East and survived for less than sixty years. In feudal fashion, the victors divided the conquered state among themselves. Baldwin of Flanders was proclaimed emperor of Constantinople (1204-5) and a Venetian, Pier Morosini, was raised to Latin (Catholic) patriarch of the East. Montferrat was granted the Kingdom of Thessaloniki and the rest of the territorial spoils were distributed as the new emperor's vassal holdings to various warriors. Dandolo and Venice received: control of a part of Constantinople; most of the Aegean islands; a portion of Achaia; Crete; and possessions in the Adriatic (including Dubrovnik, Durrës, and Zadar). For as long as it existed, the Latin Empire was progressively weakened by feudal rivalries among its constituent territorial lords and vassals.

Three Orthodox states quickly emerged as the Latins' contenders for reestablishing the stricken Byzantine Empire. The first was the so-called Empire of Nicæa, founded by refugees from Constantinople and headed by the ejected Emperor Alexius V Doukas (1204), which was established in Anatolia. Under their emperor Theodore I Laskaris (1204-22), the Nicæans effectively prevented the Latins from gaining any permanent foothold in that Asian region.

In the Balkans, newly restored Bulgaria emerged as a leading contender for the mantle of Orthodox imperial reconquest. Its ruler was Kaloyan, who defeated Baldwin and his Latin forces at Adrianople in 1205, capturing the vanquished ruler and holding him captive until his death at Tŭrnovo. Bulgaria then expanded south and west at the expense of the Latin Empire. Kaloyan toyed with Catholicism, winning papal recognition of an autonomous Bulgarian church and of himself as king. In 1218 his son, Ivan II Asen (1218-41), overthrew a usurper of the Bulgarian throne and initiated the apogee of the second Bulgarian state. Though personally mild and generous, Ivan proved an effective military commander and statesman. He further undermined the Latin presence in the Balkans, forged alliances with Nicæa, definitively broke with Rome (1232), won recognition of an independent Bulgarian patriarchate of Tŭrnovo from the Greek church, and crippled the third major contender for a Byzantine restoration, the Despotate of Epiros.

Epiros was founded by Michael Angelos Komnenos (1204-14) and expanded along the western Balkan Adriatic coastline at Latin, Venetian, and Bulgarian expense under Theodore Doukas Angelos (1214-30), who ended the Kingdom of Thessaloniki in 1222, proclaimed himself emperor, and defeated a Nicæan force near Adrianople (1224). Defeat and capture by the Bulgarians snuffed out his successes, and the influence of Epiros in the struggle for Byzantine restoration was reduced thereafter. (See Map 19.)

BOHEMIA

Danube R.

GALICIA

KIEVAN
RUSSIA

HOLY
ROMAN
EMPIRE

Vienna

Esztergom

HUNGARY

TRANSYLVANIA

Zagreb

SLAVONIA

Venice

To
Venice

CROATIA

BOSNIA

Belgrade

Zadar

PAPAL
STATES

Dubrovnik

Raška

SERBIA

BULGARIA

Danube R.

Tŭrnovo

Sofia

KINGDOM OF THE
TWO SICILIES

Naples

To
Venice

Durrës

Ohrid

Adrianople

EPIROS

K. OF THESSALONIKI

Thessaloniki

Constantinople

Nicæa

Arta

EMPIRE OF
NICÆA

D. OF ATHENS

EUBŒA

PR. OF
ACHAIA

Athens

SELJUK
EMPIRE

Mystras

RHODES

MILES

| 0 | 50 | 100 | 150 | 200 |

| 0 | 100 | 200 | 300 |

KILOMETERS

CRETE

The Latin Empire

Venetian island possessions

Holy Roman Empire

Map 18: The Rise of Serbia, 13th–14th Centuries

In the mid-11th century the Serbs of the Zeta (Montenegrin) region of the Byzantine Empire managed to establish a modicum of independence from Byzantium, and in 1077 their ruler Mihajlo (1051-81) was granted a royal crown in the pope's effort to gain a permanent foothold in the Orthodox East following the Great Schism of 1054. But the Zeta Serb tribal kingdom, isolated in its mountainous environment, was of little consequence in the Byzantine Balkans. A truly influential Serbian state emerged in Raška under the reign of Stefan I Nemanja, who after 1180 managed to throw off direct Byzantine control (1190) and unite Zeta, northern Albania, and much of present-day eastern Serbia under his authority. A devout Orthodox Christian, Nemanja definitively planted his Serb state in the East European Orthodox East. After abdicating in favor of his son Stefan II Nemanja (1196-1227), the true founder of the Nemanja dynasty, the former ruler retired first to a Serbian monastery and then to Mount Athos, where he joined another son, Rastko (known as St. Sava), in founding the large and influential Slav monastery of Hilandar.

The reign of Stefan II began amidst conflict with his brother Vukan, prince of Zeta, whom the Hungarians, pushing southward into the Balkans, supported. Stefan was forced to flee to Bulgaria, where King Kaloyan supplied him with a Cuman army in return for territories around Belgrade and Niš. The struggle with the Hungarians ended in successful mediation by Stefan's saintly brother, Sava, and Stefan assumed the throne of Serbia proper. He received a royal crown from the pope through a legate in 1217, thus acquiring the title of "First Crowned." But Sava managed to gain the recognition of an autocephalous Serbian Orthodox archbishopric from the Greek patriarch, then in Nicæa. Sava, the first Serbian archbishop, then crowned Stefan Orthodox ruler of Serbia in 1219 with a crown sent by the Nicæan patriarch, an act that ended Catholic hopes to dominate the Serbs.

Stefan's successors during most of the 13th century proved weak rulers unable to thwart the territorial ambitions of Serbia's powerful neighbors. (See Map 19.) Bulgaria entrenched its hold over Serb lands in the east, while Hungary held the region around Belgrade and established suzerainty over Bosnia. A strong secular and clerical Serbian aristocracy, which dominated the royal office, emerged during that time. Vague legal notions of succession and inheritance led to dynastic conflicts and regional disturbances exacerbated by continuing struggles between adherents of Catholicism and Orthodoxy, and by the presence of Bogomil heretics, particularly in Bosnia.

Stefan Uroš II Milutin (1282-1321), a pious yet dissolute person and an opportunist in religious and political matters, restored royal authority in Serbia. Taking full advantage of the growing weakness of the restored but truncated Byzantine Empire, Uroš gradually expanded Serbia into northern areas of Macedonia, along the Adriatic coast, and into Hungarian-held territories near Belgrade. His character was aptly demonstrated in his domestic life and in dealings with the Byzantines. He had a legal wife but maintained affairs with two known, officially kept concubines, as well as with a Greek princess from Thessaly. When his wife died (1297), Emperor Andronicus II (1282-1328) proposed a marriage alliance with Uroš, which flattered the Serb king immensely, even though he was conducting a known affair with a nun, who was also his sister-in-law. The bride was five-year-old Princess Simonis, the emperor's daughter. (Uroš was then in his forties.) They were wed in Thessaloniki in 1299. Simonis was kept in the royal nursery for a few years before the lecherous Uroš consummated the marriage.

Uroš was succeeded by his illegitimate son, Stefan Uroš III Dečanski (1321-31), who won a decisive victory over the Bulgarians and the Byzantines outside Kyustendil (Velbuzhd) in 1330. Soon afterward, Byzantium disintegrated into civil war and Bulgaria was reduced to being Serbia's subordinate ally. (See Map 20.)

Dečanski's son, Stefan Uroš IV Dušan (1331-55), brought Serbia to the pinnacle of its historical power and glory. He began his career by deposing his father and having him strangled soon afterward. He came to rule over a Serbian state that included Raška, Zeta, Macedonia, Albania, Epiros, and Thessaly down to the Gulf of Corinth. He pushed the Hungarians north of the Danube and incorporated Belgrade and its environs into his large Balkan Serbian state. His attempts to conquer Bosnia proved unsuccessful, but Dušan cemented an alliance with the Bulgarians that spread Serb influence into the eastern Balkans, and he tried to remain on fairly good terms with the Hungarians and Dubrovnik to give himself a free hand in exploiting the continuing civil disorders in the Byzantine Empire.

At his Macedonian capital of Skopje in 1346, Dušan had himself proclaimed emperor of the Serbs, Greeks (Byzantines), Bulgarians, and Albanians and was crowned as such by the archbishop of Peć, whom he then raised to the position of independent Serbian patriarch. A legal code for his Serbian "empire" was promulgated, and Dušan's court at Skopje took on all the outward trappings of Byzantine splendor. Before he could execute a planned advance on Constantinople and make good on his imperial pretensions, Dušan died. His death in 1355 was a catastrophe for the Orthodox Balkans, since it removed the last force capable of withstanding the advance into Southeastern Europe of the militantly Islamic and expansionary Ottoman Turks. (See Map 22.) Soon after his death, Serbia fell into internal disarray, with local rulers throwing off the central authority of the weakened successors to the Serbian throne.

POLAND

KIEV

GOLDEN HORDE

Danube R.

HOLY
ROMAN
EMPIRE

HUNGARY

Vienna
Bratislava
Buda

Iaşi

MOLDAVIA

Zagreb
SLAVONIA
CROATIA
BOSNIA

Cluj

TRANSYLVANIA

Venice

PAPAL
STATES

Zadar
Split

Belgrade

Tîrgovişte

WALLACHIA

Danube R.

ZETA
Dubrovnik

Ra̧ška
Peć
Duklja

Niš

Vidin

Tŭrnovo

BULGARIA

KINGDOM
OF NAPLES

Naples

Kyustendil
Skopje
Ohrid

Sofia

Plovdiv

BYZANTINE

ALBANIA

MACEDONIA

THRACE

Constantinople

Thessaloniki

EMPIRE

Butrint

THESSALY

EPIROS

To Byzantium

SICILY

Vonitsa

EUBEA

SELJUK
EMPIRE

Athens

MOREA

Mystras

	Zeta, 11th Century
	Serbia ca. 1250
	Added to Serbia in 1321
	Added to Serbia in 1355
	Added in 1321; Lost in 1355

MILES
0   50   100   150   200

0   100   200   300
KILOMETERS

CRETE
To Venice

Map 19: Eastern Europe, Mid–13th Century

By the mid-13th century, the Latin Empire was in its death throes. Undermined internally by divisive western feudal rivalries and battered externally by the constant pressures of the three primary contenders for the former Orthodox Byzantine Empire—Nicæa, Bulgaria, and Epiros—the Latin Empire, under its last ruler, Emperor Baldwin II (1228-61), consisted of little more than the European and Anatolian environs of Constantinople. The Duchy of Athens and the Principality of Achaia still remained in Latin hands. Venice retained its hold on Crete, most of the Aegean islands, and various enclaves along the Balkan Adriatic coast, including Dubrovnik.

Of the three contenders for the Byzantine patrimony, Nicæa emerged as the ultimate victor. Its rise was the result of a long and difficult process that first required consolidation of its authority in Anatolia before it could cross over into Europe and deal with Bulgaria and Epiros. In 1241 Nicæa was aided in its efforts against Bulgaria by a devastating incursion of the Mongols of the Golden Horde into that state. Later, in 1246, the elevation of a child, Mihail Asen (1246-57), to the Bulgarian throne was exploited by Nicæan emperor John Vatatzes (1222-54), who conquered most Bulgarian territory south of the Balkan Mountains, while Epiros despot Michael II (1236-71) expropriated western Bulgarian lands. Vatatzes's capture of Thessaloniki and military successes against Michael II resulted in Epiros's accepting Nicæan suzerainty (1254). Vatatzes's successor, Theodore II Laskaris (1254-58), continued the military efforts against Bulgaria but lost Epiros to a revolt by Michael II. When Laskaris was succeeded by a child, John IV Laskaris (1258-61), a military revolt led to the regency of Michael VIII Palaiologos (1259-82), who then imprisoned and blinded John in 1261.

Michael VIII forged an alliance with Bulgaria and concluded a treaty with Genoa, Venice's chief rival for commercial dominance in the eastern Mediterranean, granting it privileges similar to those enjoyed by the Venetians in the former Byzantine Empire. In July 1261 a Nicæan army crossed the Bosphorus and marched on Constantinople. Sympathetic supporters inside the city notified the army of an undefended entrance in the land walls. A small detachment of troops entered and opened a main military gate for the rest of the army, and the Latin Empire came to an end.

North of Bulgaria and Epiros, the young state of Serbia experienced growing pains. Following the death of Stefan II Nemanja in 1227, the royal crown passed first to his weak son Stefan Radoslav (1227-34), who was overthrown by his brother Stefan Vladislav (1234-42). Vladislav married a daughter of the Bulgarian tsar Ivan II Asen and Serbia fell under Bulgarian domination—especially its eastern territories. His brother and successor, Stefan Uroš I (1242-76), married a daughter of the deposed Latin emperor Baldwin II and concluded an alliance with Charles de Anjou, king of the Two Sicilies (1262-85), brother of French king Louis IX, and heir to Latin claims on the throne of Constantinople. Uroš's pro-Western ties failed to deflect Hungarian expansion into northern Serbia and Bosnia. (See Map 18.)

In Hungary, the Golden Bull, which granted privileges to an emerging aristocracy, was followed by King András's royal charter (1224) that gave the Transylvanian Saxons special privileges, including self-government under direct royal oversight. (See Map 13.) Despite the efforts of King Béla IV to reestablish a strong central authority, the now powerful Magyar landowners-functionaries continued their resistance against the throne. Béla attempted to stifle them by enlisting the support of the Cumans, whom he allowed to settle in the central plain of Hungary. Obsessed with their internal dissensions, the Hungarians were taken by surprise when the Mongols of the Golden Horde swept into the state in 1241 and defeated Béla, who fled to the Croatian Adriatic coast with the victors in pursuit. Hungary was then suddenly saved by news of the Great Han's death in faraway Mongolia. The Mongol leaders, required to participate personally in electing a successor, were forced to turn their forces back to Asia. Béla returned to a state left devastated by their ravages. The rest of his reign was occupied with wars against Holy Roman emperor Frederick II Babenberg (1212-50) and Bohemian king Otakar II Přemsyl (1253-78).

King Otakar I Přemsyl (1197-1230) made Bohemia, one of the seven imperial electors in the Holy Roman Empire since the early 12th century, a decisive force in German political affairs by playing off the various sides in the empire's early 13th-century succession struggles. He, however, proved unable to keep the Catholic church organization in Bohemia under royal control. In 1212 Emperor Frederick granted the Bohemian nobility the right to elect their rulers. Otakar's successor, Václav I Přemsyl (1230-53), invited numerous German immigrants into Bohemia to counteract the Czech nobility's growing power. The Germans received numerous royal privileges and did much to expand commerce and agriculture. The political power forged by Otakar I and the economic prosperity begun under Václav were brought to full fruition during the reign of Otakar II.

In Poland, the first half of the 13th century was marked by near total feudal anarchy, dizzying dynastic struggles over the throne, and rising political free-agency among the aristocracy and clergy. In the midst of the turmoil, Poland was ravaged by two Mongol invasions (1241 and 1259) and forced to accept the arrival of the Teutonic Knights (1228), who were invited in as vassal allies to deal with the pagan Prussians in the north, but who chose to build a state for themselves once the job was completed. Their state served as a barrier between Poland and the Baltic Sea.

Map 20: Eastern Europe, Mid–14th Century

The death of King Václav III (1305-6) ended the Přemsyl dynasty and the Czech kings' hold on the Polish throne, which had begun in 1290 with Václav II (1278-1305). An interregnum followed in Bohemia that ended when the Czech nobility elected as king John of Luxemburg (1310-46), who was forced to grant them a charter guaranteeing their rights and privileges. A permanent Bohemian diet was established. John, a foreigner, spent little time in his kingdom. Instead, he resided mostly in Paris or wandered around Europe, fighting with the Teutonic Knights against the rising state of Lithuania or offering his services as a skilled warrior to France in the Hundred Years War. He became blind in his later years, yet he fought on, dying while charging the English in the Battle of Crécy (1346).

John's son, Charles I Luxemburg (1346-78), instituted the "golden age" of Bohemia. Born and partly raised there, Charles thought of himself as Bohemian. Shortly after becoming king, he was elected Holy Roman emperor (1347) as Charles IV. Although the Germans considered him a poor emperor, the Czechs viewed him as a great king. He established a stable system of succession, and his Golden Bull of 1356 gave the office of Bohemian king first position among the empire's secular electors. Charles governed essentially as a constitutional monarch, and a new law code, the *Maiestas Carolina,* was issued. Silesia and Brandenburg were incorporated into Bohemia.

Perhaps most significant for the Czechs, Charles's capital at Prague was made the center of the whole Holy Roman Empire and transformed by him into the cultural capital of Central-Eastern Europe. In 1348 he founded there the first university east of the Rhine River, a school that he named after himself, and one of the oldest such institutions in all of Europe. The city was extensively rebuilt and beautified in the latest Gothic architectural style with numerous palaces and fortifications. Arts and letters found rich patronage from the emperor-king. (See Map 21.)

Elsewhere in Northeastern Europe, the Teutonic Knights consolidated their hold over most of the Baltic coastline, and from their capital at Malbork (Marienburg) placed steady pressure on Poland to their south. Numerous Germans settled there, especially in the Polish towns, which (to the chagrin of the Polish rulers) essentially became German-speaking. To their east, the Knights were embroiled in a deadly conflict with the large, still highly pagan, emerging state of Lithuania. (See Map 23.)

In Poland, succession problems ended with the reign of Władisław IV Łokietek (1305-33). Close contacts with Bohemia continued. Charles of Bohemia represented a powerful role model for Łokietek's successor, King Casimir III the Great (1333-70), who imitated Charles by establishing a university in 1364, the second oldest in Eastern Europe, in his capital at Cracow. He founded the school in an effort to centralize higher education in Poland and to institutionalize Polish cultural development. Cracow thereafter represented not only Poland's political capital but its cultural fountainhead as well. At Casimir's death, the throne passed to the foreigner Louis of Anjou (1370-82).

The end of the Árpád dynasty in Hungary (1301) was followed by a period of dynastic turmoil in which the feudal magnates, high Catholic clergy, and gentry entrenched their privileges against weak foreign monarchs. The French-Sicilian house of Anjou gained control of the throne with Charles I Robert (1308-42), who introduced Western influences such as chivalry. He gradually reduced the magnates' power, brought the nobility under royal regulation, and obliged them to serve in the state military. Charles also instituted the first direct state tax and encouraged urban development. He succeeded in entrenching royal power within the traditional feudal order that held in Hungary. His son, King Louis I the Great (1342-82), established court in Buda, where he patronized learning and culture. Louis placed further curbs on the magnates. Attempts to solidify Angevin interests in Naples brought him into conflict with Venice (1347-81), resulting in the acquisition of the Croatian Dalmatian coast. In 1370 he was elected king of Poland. Louis exerted his influence deep into the Balkans—Bosnia, Wallachia, and newly risen Moldavia all recognized Hungarian suzerainty—and he founded Balkan frontier defensive districts against the advancing Ottoman Turks. (See Map 22.)

In the Balkans, Serbia at mid-century reached its greatest territorial extent under Dušan, but was about to slip into rapid internal decline following his death in 1355. (See Map 18.) Bulgaria had already done so.

From its crushing defeat by the Serbs in 1330 and the death of Tsar Mihail Shishman (1323-30) until the death of Dušan, Bulgaria essentially existed as a Serbian satellite state. Although Tsar Ivan AleksandŭrAleksandŭr (1331-71) enjoyed a long reign, the authority of the Bulgarian imperial office experienced a sharp decline, while the power of regional magnates rose at its expense. By 1365 the state was divided into feudal principalities, the most viable of which were Bulgaria proper, centered on the traditional capital at Tŭrnovo, and Vidin, ruled by Ivan Stratsimir (1365-96).

Byzantium's state throughout the first half of the 14th century was chaotic. The Catalans, Spanish mercenaries brought in to fight the Turks in Anatolia but who then turned on the Byzantines, ravaged the empire's European possessions (1305-11) and took control of the Duchy of Athens. A civil war followed (1321-28) between Andronicus II (1282-1328) and Andronicus III (1328-41), and another (1341-47) between John V Palaiologos (1341-76) and his co-emperor John VI Kantakuzenos (1347-54). During this time, Dušan meddled successfully in Byzantine affairs and the Ottoman Turks won a foothold in Europe. (See Map 22.)

SMOLENSK

Vilnius

Danzig
Malbork

TEUTONIC KNIGHTS

Minsk

LITHUANIA

Berlin

BRANDENBURG

Gniezno

HOLY
ROMAN
EMPIRE

SILESIA

POLAND

K. OF BOHEMIA

Cracow

KIEV

Kiev

Prague

L'viv

GALICIA

Danube R.

Augsburg

SLOVAKIA

Košice

AUSTRIA Vienna

Bratislava

MOLDAVIA

GOLDEN
HORDE

STYRIA

Buda

Cetatea
Albă

CARNIOLA

Zagreb

HUNGARY

Cluj

TRANSYLVANIA

Venice

Danube R.

CROATIA

To Venice Zadar

BOSNIA

Belgrade

WALLACHIA

PAPAL
STATES

Split

HERCEGOVINA

Danube R.

Vidin

Tǔrnovo

ZETA

Peć

Rome

SERBIA

Sofia

BULGARIA

Naples

Durrës

To
Venice

Skopje

BYZANTINE
EMPIRE

Constantinople

K. OF
NAPLES

EPIROS

Thessaloniki

Gallipoli

Bursa

To
Byzantium

OTTOMAN
EMPIRE

K. OF
SICILY

D. OF
ATHENS

Athens

SELJUK
EMPIRE

ACHAIA

Mystras

DESPOTATE
OF THE
MOREA

RHODES

To Byzantium

To Venice

CRETE

MILES
0 50 100 150 200

0 100 200 300
KILOMETERS

—— Eastern border
of the Holy Roman Empire

☐ Kingdom of Bohemia

Map 21: Prague, Mid-14th–15th Centuries

Prague developed prior to the 9th century out of a cluster of villages perched on hills that overlooked a ford of the Vltava River (a tributary of the Elbe) used by overland merchants and traders. In the early 9th century the hills were fortified by the Přemsyls (Czech tribal leaders) and slowly grew into their administrative and cultural center. By the mid-10th century a small Jewish merchant community was established across the Vltava from the Czech fortifications on Hradčany Hill, and the town had acquired an international reputation as a thriving trading and manufacturing center. At that time, Hradčany sported two stone churches, St. George and St. Vitus, erected by the Přemsyl princes Vratislav (905-21) and St. Václav, respectively. The rest of the town was constructed of wood and mud.

Prince Boleslav II (967-99) established a Catholic bishopric at Prague in 973, with the church of St. Vitus as its cathedral. Boleslav's action removed the Czechs from the German bishopric of Regensburg's authority. Soon a Benedictine monastery was founded on the town's outskirts and a Benedictine convent was attached to the church of St. George. Prague thus became the vanguard in the Přemsyls' attempts to elevate their territories both spiritually and culturally.

When Vratislav II (1085-92) was crowned the first king of Bohemia in 1085, he established his court in Prague, and a new basilica of St. Vitus was built to replace the older rotunda structure. The city flourished as a wealthy center of trade, which was concentrated in the area of Old Town Square, and rich merchants erected stone houses around the marketplace. German traders came to predominate during the reign of Vladislav II—since he was a staunch ally of Holy Roman emperor Frederick I Barbarossa—and they founded an independent colony in the town. Trade made possible urban expansion and renovation. Hradčany Castle was transformed into a Romanesque palace, and old St. George Church was turned into a new basilica. A stone bridge was thrown across the Vltava, linking the castle to the trading district, becoming only the third such structure in the central part of Europe. A Premonstratensian monastery was founded at Strahov. By the end of the 12th century, Prague ranked among the largest towns in the Holy Roman Empire, containing some fifty churches and numerous stone houses of powerful royal courtiers and wealthy merchants.

During the reign of King Václav I, around 1230, Prague's castle district, Small Quarter, and Old Town were enclosed within walls and the inhabitants granted a charter of rights and freedoms. Known as "German Law" because it was based on models granted German urban colonists in old medieval towns in northern Europe, the Old Prague Law provided for freedom of economic activity and a certain amount of self-government (an elected town council and an independent criminal court, among other rights). The promulgation of such urban law gave Prague official recognition as a city.

Throughout the 13th century, Western culture flowed into Prague. Gothic art and architecture augmented, and oftentimes replaced, the older Romanesque style; German and Italian customs and fashions became popular to Prague's inhabitants, from the court downward. This trend reached its height during the reign of King-Emperor Charles I Luxemburg. (See Map 20.)

Although he spent considerable time as a youth at the French court, Charles was strongly attached to Bohemia and Prague. He lavished much attention and patronage on the city, which served as both his royal and imperial capital. He renovated Hradčany Castle à la those of the French kings, demolished Romanesque St. Vitus basilica and replaced it with a Gothic cathedral in honor of the Prague bishop's elevation to archbishop (1344), and founded Prague University (1348), the fourth such institution in Europe after those in Bologna, Padua, and Paris. He then expanded the city by building the New Town, in the latest design of European urban planning, to free the university from the noise and hubbub of the trading conducted in the Old Town. The New Town was designed to be an ideal city, with well-ordered and symmetrical squares and streets, uniformly tall houses, and towering religious complexes placed at visual focal points. Slavic culture was cultivated through the Benedictine monastery of Emmaus, which had papal permission to practice the Catholic liturgy in Old Church Slavonic. At Charles's death (1378), Prague was the third largest city in Europe, after Constantinople and Rome.

Charles's son, Václav IV (1378-1419), inherited his father's cultural interests. Though a failure politically, he continued to endow Prague with royal patronage. He also demonstrated a preference for the commoners of his kingdom over the aristocracy by moving his residence from Hradčany Palace to the Old Town. A new building, the Carolinum, was given to the university in honor of his father, and Václav was close friends with the school's rector, Jan Hus, who preached in the university's Bethlehem Chapel both against the Catholic church's immorality and for changing social and ethnic conditions in Bohemia and throughout Europe in general. During Václav's reign, Prague became the epicenter of the Hussite movement that rocked the Holy Roman Empire and and the Catholic church during the first half of the 15th century and that later served as the early foundation of Czech nationalism. With the rise of the Hussites, Prague lost its position as capital of the Holy Roman Empire.

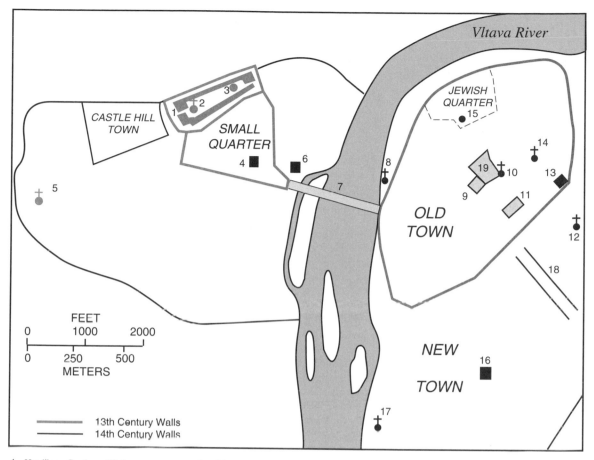

Vltava River

CASTLE HILL TOWN

SMALL QUARTER

JEWISH QUARTER

OLD TOWN

NEW TOWN

FEET

| 0 | 1000 | 2000 |

| 0 | 250 | 500 |

METERS

13th Century Walls
14th Century Walls

1. Hradčany Castle and Palace
2. St. Vitus Cathedral
3. Church of St. George
4. Small Quarter Town Hall
5. Strahov Monastery
6. Bishop's Residence
7. Charles Bridge

8. Monastery of the Knights of the Cross
9. Old Town Hall and Clock Tower
10. Church of the Tyn
11. Charles University, including the
    Bethlehem Chapel and the Carolinum
12. Church of St. Henry
13. Powder Tower

14. Church of St. James
15. The Old-New Synagogue
16. New Town Hall
17. Benedictine Abbey of Emmaus
18. Horse Market
19. Market Square

Map 22: The Rise of the Ottoman Empire, 13th–15th Centuries

The term "Ottoman" is a Western corruption of the Turkish name of their original tribal leader, Osman I (1281-1324). He ruled the Seljuk principality closest to Byzantium and Europe in the northwest corner of Anatolia, and pursued unrelenting warfare against the Christians directly across his borders. This soon attracted to him a number of warriors—organized into an effective and loyal military force—from all parts of the Seljuk world eager to expand the Islamic territories in the tradition of the *jihad*. After his death, Bursa was captured and the Byzantines were completely expelled from Anatolia by his son and immediate successor, Orhan I (1324-60). Under Orhan, the Ottomans (the collective name for the assorted warriors and allies of the house of Osman) permanently established themselves in Southeastern Europe. For the next two-and-a-half centuries, their military successes against European Christians stretched in an unbroken string under ten consecutive rulers.

Ottoman forces first entered Europe in 1345 as mercenary allies of John VI Kantakuzenos in his civil war with Emperor John V Palaiologos. In 1349 Orhan again sent military support to John VI to counter Dušan's Serbian encroachments. When John called on Turkish help for the third time in 1354, Orhan's forces did not return to Anatolia as they had previously, but took control of Gallipoli in Europe, fortified it, and transformed it into a permanent base for expansionary operations in Southeastern Europe. By Orhan's death, the Ottomans were entrenched in Europe, their state was well-organized, and the Byzantine Empire was at their mercy.

Murad I (1360-89) realized that the Balkan Christian states of Byzantium, Serbia, and Bulgaria had been weakened by decades of internecine wars. Their populations were burdened with rising semifeudal oppression, economic disruption, and unstable living conditions. Murad captured Adrianople in 1365 and transformed it into his European capital, Edirne. A new standing professional infantry force, called the Janissaries, was founded to supplement the traditional Turkish tribal cavalry units. The new troops consisted first of enslaved war captives and later of child-levies from among the growing numbers of Balkan subject Christians. Murad conquered the Bulgarian lands south of the Balkan Mountains by 1372 and reduced its Tŭrnovo ruler, Tsar Ivan Shishman (1371-93), to vassal status. In 1371 he destroyed a predominantly Serbian force outside Edirne, an act that resulted in the conquest of Macedonia, and by 1386 he had taken the regions of Sofia and Niš, and forced a weakened Serb ruler, Prince Lazar (1371-89), into submission as his vassal.

As Murad's state expanded, it became clear that the Ottomans were in Europe to stay. Many Christian rulers, such as Shishman and Lazar, as well as their independent-minded warrior nobility, joined the Ottomans as allies in an effort to retain their political and social positions. Christian Bulgarians and Serbs fought loyally in the Ottoman ranks throughout most of the campaigns that won control of the Balkans, including that which resulted in the Battle of Kosovo Polje (1389), when Serbia was broken definitively. Lazar had joined a coalition of Serb, Bulgarian, Bosnian, Albanian, and Wallachian magnates in an effort to contain the Turks. Murad set out to punish his unfaithful vassal. In the battle, Lazar was killed and Murad assassinated by a Serb pretending to have defected.

Murad's death did little to stop the Turkish onslaught. Bayezid I the Thunderbolt (1389-1402) continued Ottoman expansion by rapid campaigns that made Serbia a vassal state, incorporated Bulgaria outright (1393), and reduced Wallachia to subordinate vassal status. Constantinople was unsuccessfully besieged (1391-98) and much of Greece conquered. A Western crusade against the Turks, led by Hungarian king Sigismund (1387-1437), was destroyed near Nikopol on the Danube in 1396. But Ottoman European momentum was stymied by the Mongol invasion of Anatolia (1402), which Bayezid was powerless to stop. There followed his death an interregnum (1402-13), during which Ottoman expansion in the Balkans temporarily stalled and Serbia was able to reassert a modicum of independent action.

The brief respite for the Balkan Christians ended when the Turks returned in renewed force under Mehmed I the Restorer (1413-21), who reasserted his authority over most of the Turks' European possessions. His son, Murad II (1421-51), resumed Ottoman expansion. A war with Venice resulted in the conquest of Thessaloniki (1430) and most of the Aegean islands. Hungary was unsuccessfully invaded (1442) and a subsequent Hungarian-led crusade, the last against Islam, was crushed outside the Bulgarian seaport of Varna (1444). At Murad's death, little remained of the Byzantine Empire other than Constantinople itself.

Mehmed II the Conqueror (1451-81) assumed the throne determined to capture the imperial city and transform it into the capital of an Islamic Ottoman Empire. After cutting the city off from all outside assistance, Mehmed laid siege to Constantinople in 1453. His artillery blasted holes in the once impregnable land walls, and on 29 May the final vestige of the Roman Empire fell with its last emperor, Constantine XI (1448-53), dying heroically defending the walls. The city was renamed Istanbul and repopulated with Turks, Greeks, Armenians, and assorted Balkan Christians.

Once established in his new capital, Mehmed set out to finish the total conquest of the Balkans and to push the borders of his empire deeper into Europe. Serbia was finally subdued and incorporated (1456-58), Bosnia-Hercegovina conquered (1458-61), the tough resistance of the Albanians, led by George Kastriotis/Skanderbeg, broken (1456-63), and Venice's presence in the eastern Mediterranean reduced (1463-79).

POLAND

Danube R.

Vienna

HOLY
ROMAN
EMPIRE

Buda

Cluj

Suceava

BESSARABIA

MOLDAVIA
(Vassal Client)

HUNGARY

TRANSYLVANIA

Cetatea Albă

VENICE

Zagreb

Venice

CROATIA

SLAVONIA

Belgrade

Tirgovişte

Bucharest

PAPAL
STATES

Zadar

DALMATIA

BOSNIA

Smederevo

WALLACHIA
(Vassal Client)

Sarajevo

SERBIA

Vidin

Nikopol

Danube R.

DOBRUDZHA

HERCEGOVINA

Dubrovnik

MONTENEGRO

Kosovo
Polje

Niš

BULGARIA

Türnovo

Varna

Naples

Lech

ALBANIA

Skopje

Sofia

NAPLES

Durrës

Ohrid

MACEDONIA

Plovdiv

Adrianople

THRACE

Thessaloniki

Constantinople

Joannina

EPIROS

Larissa

Gallipoli

Bursa

MILES

THESSALY

BYZANTINE
EMPIRE

0   50   100 150 200

Navpaktos

EUBOEA

Athens

SELJUK
PRINCIPALITIES

0   100   200   300
KILOMETERS

(To
Genoa)

K. of
SICILY

Mystras

MOREA

Methoni

Monemvasia

RHODES
(To
Knights of
St. John)

(To Venice)

CRETE

	Venetian possessions
	Ottoman possessions,1326
	Territories acquired by 1360
	Territories acquired by 1389
	Territories acquired by 1402
	Territories acquired by 1481
	Territories reduced to vassal clientage by 1402, permanently annexed by 1481
	Territories acquired by 1504

Map 23: The Expansion of Poland, 14th–15th Centuries

Although political stability and cultural flowering characterized Poland during the 14th century, the Teutonic Knights and a growing German settler population from the northern Baltic shore regions caused continuing difficulties for the Polish state. The German presence threatened Poland's hold on Pomerania and, by extension, its access to the Baltic Sea. The Poles proved unable to effectively handle the increasingly independent Teutonic Knights, and violent clashes between Poles and Knights intensified throughout the 14th century.

All the while, the Piast dynasty was weakening. At Casimir the Great's death, the throne passed to the foreigner Louis of Anjou (1370-82), whose reign, in turn, was followed by a period of civil strife until in 1384 the Princess Jadwiga, Louis's daughter, succeeded to the throne as the "Maiden-King" (1384-86). Partly out of desperation in facing the dangers posed by the Teutonic Knights along the shores of the Baltic, in 1386 Jadwiga married Władisław V Jagiełło, prince of Lithuania, who, like the Poles, found his state threatened by the Knights' militant presence.

Lithuania at the time was a vast, highly Russianized state whose primary interest lay in confronting Muscovy in the east as successor to Kiev. Most of Lithuania's subjects were Orthodox Slavs who were permitted to retain their identity by their Lithuanian lords, many of whom themselves had forsaken paganism and converted to Orthodoxy. Lithuania's primary common interest with Poland was the threat of the Teutonic Knights' cutting off access to the Baltic Sea.

Jagiełło converted to Catholicism and assumed the Polish throne (1386-1434), thus uniting the states of Poland and Lithuania through his status as common ruler. In 1410 the combined Polish-Lithuanian forces decisively defeated the Knights in the Battle of Grünwald (Tannenberg) and brought Poland brief domination over the Baltic shore. Although the German threat to Poland lessened after Grünwald—which had resulted in Prussia's submission to Polish suzerainty—Teutonic Prussia remained in existence as an autonomous ducal principality.

Events that occurred between the death of Casimir the Great and the union of Poland with Lithuania had a lasting impact on the future fate of Poland. Throughout that period, royal power and authority dwindled while that of the Polish nobility grew. Lacking male heirs, Casimir forged an agreement with his nephew, Hungarian king Louis of Anjou, authorizing Louis's succession to the Polish throne in return for his guaranteeing the nobility's privileges. Casimir's action instituted the elective principle in Poland that gave the nobility the right to choose their kings. The nobles, taking advantage of the lack of Piast male progeny, exacted further concessions from Louis, who had no male heirs of his own. To ensure their election, Jadwiga and Jagiełło relinquished additional royal privileges to the increasingly dominant nobility. By the 15th century, the Polish royal office was greatly limited in its power and especially restricted in its authority relative to the noble class, which had gained the sole right to elect the ruler, exacted legal recognition of its status as a closed political entity, won numerous tax exemptions, and removed the ruler's central control over the noble-led Polish military.

Jagiełło's situation in Poland lay in stark contrast to his position as grand prince in Lithuania. There rule was hereditary and the nobles, though powerful, were bound to central authority. Only Jagiełło's reputation as a strong ruler, who could handle the Teutonic Knights and would further Poland's position in the Baltic and western Ukrainian regions, led the independent-minded Polish nobles to accept him. They saw a strong Poland under Jagiełło as a boon for their dominance within the state. In uniting Poland with Lithuania, which at the time was three times larger than Poland and controlled vast territories in Ukraine on the steppes, Jagiełło opened to the Polish nobility the prospect of playing a leading role in the single largest European state, one that stretched from the Baltic Sea in the north to the Black Sea in the south.

Despite the dynastic union, both Poland and Lithuania remained technically autonomous states until the Union of Lublin (1565). (See Map 26.) Throughout the 15th century they often quarreled over their respective borders. In general, Poland succeeded in expanding its territory at Lithuania's expense.

Fifteenth-century Jagiełłonian Poland-Lithuania emerged as a Western European Great Power that rivaled the Austrian Habsburgs, whose fortunes were also on the rise. Poland-Lithuania was, in fact, the first European state to consciously try to attain such an exalted international position. In 1440 Polish king Władisław VI Jagiełło (1434-44), Jagiełło's son and successor, won election to the Hungarian throne as László I, but after his death in the Battle of Varna against the Ottoman Turks, the Jagiełłonians lost the Hungarian crown until 1490. The son of Casimir IV Jagiełło (1447-92), Władisław VI's eventual successor, acquired the Bohemian royal crown in 1471 as Vladislav II (1471-1516) and later that of Hungary as well (1490-1516). His son Louis II (1516-26) then succeeded him on both thrones. The Jagiełłonian family's acquisition of the Bohemian and Hungarian thrones was part of a conscious expansionary effort—sometimes termed the "Jagiełłonian System" by modern historians—to counter the Habsburgs' growing power in Central-Eastern Europe. The dynasty's "system" of Great Power politics reached its height around the year 1500. By that time it was coordinated through periodic meetings in Warsaw of Jagiełłonian rulers from various parts of Europe.

SWEDEN

DENMARK

MUSCOVITE RUSSIA

Novgorod

Riga

LIVONIAN KNIGHTS

Moscow

TEUTONIC KNIGHTS

Gdańsk

Vilnius

Smolensk

POMERANIA

Gniezno

Grünwald

Minsk

MAZOVIA

LITHUANIA

Oder R.

GREAT POLAND

Vistula R.

Warsaw

Bug R.

SILESIA

Prague

HOLY ROMAN EMPIRE

LITTLE POLAND

UKRAINE

L'viv

GALICIA

Kiev

Dnieper R.

GOLDEN HORDE

Danube R.

Vienna

Esztergom

MOLDAVIA

Dniester R.

Dnieper R.

Zagreb

HUNGARY

CROATIA

Danube R.

	Poland, mid-13th Century
	Additions to Poland, late 14th Century
	Additions to Poland, mid-15th Century
	Lithuania, mid-15th Century
	Losses between mid-13th and late 14th Centuries

MILES

0   100   200   300

0   100   200   300   400

KILOMETERS

Part III

Early Modern Period

(16th–18th Centuries)

Map 24: Apex of the Ottoman Empire in Europe, Mid-16th Century

The reign of Sultan Süleyman I the Magnificent (1520-66) represented the high-water mark of Ottoman expansion in Europe. Süleyman's military forces, anchored by his highly disciplined corps of household slave Janissary infantry and guard cavalry, were victorious in most battles with their European enemies. Military successes, and the extension of his empire deep into the Danubian Basin, made Süleyman preeminent among European rulers, and his reign marked the "golden age" of Ottoman architecture, fine arts, law, literature, diplomacy, and commerce.

Son of Sultan Selim I the Grim (1513-20), who spent his reign focused on Ottoman expansion in West Asia and North Africa, Süleyman turned his attentions to Europe. In 1521 he captured Belgrade, a key fortress guarding the southern border of Hungary. Intermittent border warfare with the Hungarians culminated in the decisive Battle of Mohács (1526), in which the Hungarians were crushed and their king, Louis II Jagiełło, killed. Mohács opened the way to further Ottoman advances into the heart of the Danubian Basin. The Hungarian capital of Buda was occupied, and Süleyman, deciding at first not to annex Hungary, made it a tributary state under his Transylvanian vassal, János Zápolya (1526-40). This triggered a civil war within Hungary pitting Zápolya against the Habsburg contender for the throne, Ferdinand I (1526-64). When in 1527 Ferdinand's forces captured Buda and defeated Zápolya in battle, the Transylvanian appealed to Süleyman for help.

Süleyman prepared a new expedition into Hungary and concluded a secret anti-Habsburg alliance with France (the first between a European Great Power and the Turks). In May 1529 he led his forces north of the Danube. Buda was recaptured after a short siege in September and Süleyman pushed on into the Habsburg Empire, but stubborn Habsburg resistance and the approaching winter weather frustrated his attempt to take besieged Vienna and forced him to withdraw. Ottoman-Habsburg fighting continued in the north of Hungary until 1533, at which time a war with Persia in West Asia forced Süleyman to sign a treaty with Ferdinand recognizing Habsburg control of a strip of northern and western Hungary, while Zápolya retained his hold on the remaining two-thirds of that kingdom. Both were made tributary to Süleyman for their Hungarian possessions.

Ferdinand resumed warfare against Süleyman in 1537 as part of a joint anti-Ottoman and anti-French alliance with the Holy Roman Empire, the Papal States, and Venice organized by his brother, Holy Roman emperor Charles V (1519-56). In 1541, with Habsburg forces consistently unsuccessful and with the death of Zápolya (1540), Süleyman decided to annex Zápolya's portion of central Hungary. Three years of inconclusive campaigning followed. Events in West Asia made it necessary for Süleyman to renew the war with Persia, so he felt constrained to make peace with the Habsburgs on the European frontier. In 1544 he made another agreement with Ferdinand based on the *status quo ante*. This permitted Ferdinand to hold his strip of Royal Hungary in return for continued annual tribute payments.

A short lull in the Ottoman-Habsburg conflict followed the signing of the peace. Transylvania remained a tributary vassal state of the Turks under a native prince. The Romanian principalities of Wallachia and Moldavia were also governed by native princes who were tribute-paying vassals of the sultan. While there was no Turkish military presence in Transylvania, the Ottomans maintained a few garrisons in fortresses built on Romanian territories to protect the empire's Danubian and Ukrainian defenses. All three vassal states were required to pay the Turks an annual tribute, their ruling princes could not assume power without the confirmation of the Ottoman sultan, and certain trade commodities (especially food stuffs) had to be sent to the empire. In return, the Turks permitted them internal political, social, and cultural autonomy. So long as the vassal states met their monetary and service obligations and did not act on the international scene in ways deemed detrimental to Ottoman foreign policy, the Turks were content to let the native princes govern without much Turkish interference. Elsewhere in the Balkans, only portions of Croatia and Dalmatia lay outside the sultan's direct authority.

In 1551 Ferdinand renewed the war against the Turks by invading Transylvania, which he succeessfully held for two years before being repulsed. Following Charles V's abdication in 1556, Ferdinand was elected Holy Roman emperor (1556-64) and the war with the Ottomans settled into desultory border fighting. Süleyman and Ferdinand brought the intermittent fighting to a close by the Peace of Prague (1562) with no real change in the status quo. Ferdinand continued to pay Süleyman tribute for his Hungarian holdings until his death in 1564. When Ferdinand's imperial successor, Emperor Maximilian II (1564-76), ordered renewed raids against the Turks in 1566, Süleyman, then 72 years old and suffering from gout, led an army against Royal Hungary and laid siege to the fortress of Szigetvar. There he died in his tent, two days before the citadel fell to his troops, who were prevented from learning of his death until after their victory.

Although the Turks would continue to pose a serious threat to Western Europe for another century, and would even gain temporary new conquests in southern Poland and Ukraine, Ottoman fortunes in Europe crested with the life and death of Süleyman. The decline began during the reign of his son, whose character aptly fit his descriptive title, Sultan Selim II the Sot (1566-74).

POLAND

Danube R.

HABSBURG
EMPIRE

Vienna • • Bratislava

ROYAL HUNGARY

BESSARABIA

Suceava •

MOLDAVIA
(Vassal Client)

Buda • Pest

Cluj •

Cetatea
Alba •

HUNGARY

TRANSYLVANIA
(Loose Vassal Client)

Venice •

Zagreb •

Szigetvar •

Mohács •

SLAVONIA

WALLACHIA
(Vassal Client)

CROATIAN MILITARY BORDER

Belgrade •

CROATIA

DOBRUDZHA

PAPAL
STATES

Zadar •

DALMATIA

BOSNIA

SERBIA

Bucharest •

Danube R.

Split •

Sarajevo •

Niš •

Varna •

MONTENEGRO

BULGARIA

OTTOMAN

Dubrovnik
(Client Vassal)

KOSOVO

Sofia •

Naples •

Skopje •

Plovdiv •

KINGDOM
OF NAPLES

Durrës •

ALBANIA

Ohrid •

MACEDONIA

Edirne •

THRACE

Istanbul •

Thessaloniki •

EMPIRE

Ioannina •

K. of
SICILY

EPIROS

Larissa •

THESSALY

CORFU

Navpaktos •

MILES

Athens •

0   50   100 150 200

MOREA

0   100   200   300
KILOMETERS

Methoni •

Monemvasia •

RHODES

CRETE

	Ottoman border, mid-16th century
	Territories acquired between 1505 and 1566
	Ottoman vassal client states
	Venetian possessions

Map 25: Istanbul, 16th–17th Centuries

Sultan Mehmed II envisioned Constantinople as the capital of a powerful, highly cultured Ottoman Islamic world-state representing the divinely ordained order for all humankind on earth (a view similar to the traditional one of the Orthodox Christians regarding Byzantium). In many respects, Istanbul was to continue Constantinopolitan traditions as the political and cultural fountainhead for an essentially theocratic society. Formerly, that society had been Christian Byzantium. Now it was to be the Islamic Ottoman Empire.

After three days of obligatory sacking, as the city had been captured by force, Mehmed began embellishing his new capital. Justinian's cathedral of Hagia Sophia was converted to an imperial mosque (Aya Sofya), as eventually were other churches and monasteries. The rights of non-Turkish inhabitants were protected to ensure continuity and stability for commercial activities. Since Constantinople had never fully recovered from the sack of 1204, and because after 1261 the restored Byzantine Empire existed in a state of near-poverty, by the time of the conquest Constantinople had become a hollow shell of its former self. Its population had dwindled and much property was abandoned or in a state of disrepair.

The Turks began to repopulate the city soon after acquiring it. Civic and private properties were offered to the public by Mehmed to entice much-needed skilled artisans, craftsmen, and traders, of all religions and ethnicities, to reside within the city's walls. Istanbul rapidly grew into a multiethnic, multicultured, and bustling economic, political, and cultural center for the Ottoman state, whose distant frontiers guaranteed it peace and security.

Within its encompassing ancient walls, which the Turks permitted to fall into disrepair because of the city's secure location, Istanbul steadily acquired a distinctive Islamic character. First rose numerous minarets added to former Christian churches converted into mosques. Then came a plethora of new structures: mosque complexes, fountains, caravansaries, public baths, public soup kitchens and hospices, tombs and mausoleums, dervish convents, libraries, and other such works. The old Byzantine palaces having fallen into disrepair or ruin, Mehmed first built a new one for himself (the Eski Sarayı), scavenging materials from the older, dilapidated Church of the Holy Apostles. But he soon decided (in 1459) to construct a larger one on the crest of the old acropolis of Byzantium, in the areas of the former Byzantine Great and Mangana palaces. Surrounded by massive defense walls and encircled by extensive parks and gardens, Topkapı, which eventually consisted of a complex series of small, individual private (bedrooms, harem, libraries, and kiosks) and functional (divan, treasury, reception halls, kitchens, guard barracks, and arsenal) buildings, became the political nerve center of the empire. One of the gates in the palace walls through which all foreign plenipotentiaries exclusively treated with the imperial court eventually gave the Ottoman government

its common identity as the "Sublime Porte."

All political, military, religious, and cultural life in Istanbul revolved around Topkapı and nearby Aya Sofya, which mirrored in location their counterparts in the former Byzantine capital. Here were stationed the Janissaries; public spectacles were conducted in the Hippodrome; important religious and political functionaries built palaces in the area. Much of the loot from successful military campaigns, and a significant percentage of tribute and taxes from conquered territories, was used to adorn Istanbul with beautiful mosques, palaces, and pious foundations. No sultan was more active in such activity than Süleyman I, whose imperial architect, Sinan, gave the city some of its most exceptional masterpieces. The Süleymaniye mosque complex (1557) by Sinan, which dominated the city's skyline, was a symbol of the power and glory of the empire at its height, serving essentially the same purpose as did Hagia Sophia for Byzantium.

The inhabitants of Istanbul were a patchwork of ethnicities and religions, who often lived in city quarters identified by a particular ethnicity. There were Jewish and Armenian quarters, for example. The most significant quarter, as far as Eastern Europe was concerned, was the Phanar (lighthouse) district on the south bank of the Golden Horn near the northern terminus of the old land walls. This was the quarter inhabited by wealthy Greek merchants, and where the Greek Orthodox patriarchate was eventually established. The Greeks of the quarter, known as Phanariotes, played important economic and political roles in the Ottoman Empire in the 16th through 18th centuries, during which they dominated the empire's international maritime and overland commercial activities, commanded its navy, staffed the imperial office for foreign affairs as administrators and translators (because of their extensive foreign contacts), and purchased the crowns of the tributary Wallachian and Moldavian principalities from the Turks throughout the 18th century.

Most important, the Phanariotes controlled the office of Greek patriarch, which, because of the Ottoman *millet* system, gave them authority over all Orthodox Christians, no matter their ethnicity, within the empire. Since the state was a Muslim theocracy ruled by Islamic sacred law, non-Muslims were technically outside the law, so they had to be governed by their respective religious laws. To solve this problem, Mehmed II divided his non-Muslim subjects among Orthodox, Jewish, and Armenian *millets* (religious "nations"), whose ecclesiastical hierarchies were made responsible for administering and representing their respective members within the state. The Orthodox *millet* was the largest, representing the most significant source of imperial tax revenues, thus rendering its upper hierarchy a powerful component of the Ottoman domestic government. The patriarchate enjoyed more concrete authority within the Ottoman Balkan Orthodox world than it had previously in the Byzantine.

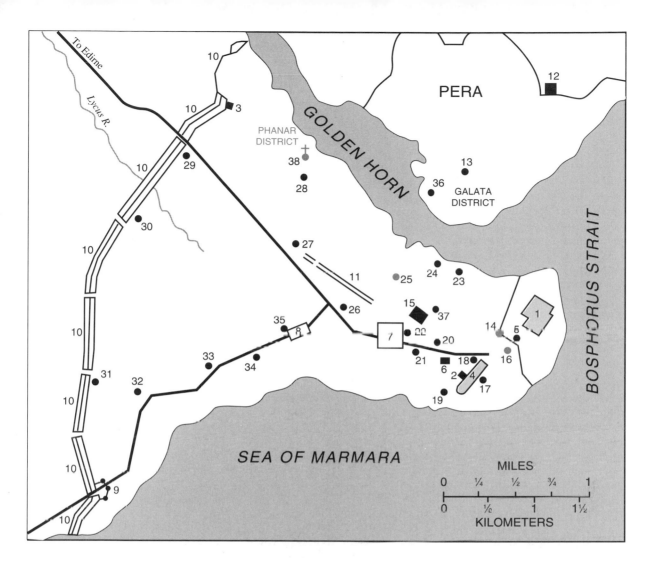

Palaces
1. Topkapı
2. Ibrahim *paşa*
3. Tekfur Saray

Civic Structures
4. Hippodrome
5. Janissary Arsenal (Old St. Irene)
6. Köprülü Complex
7. Bayezid Square
8. Aksaray Square
9. Seven Towers Fortress
10. Land Walls (partial ruins)
11. Aqueduct of Valens
12. Tophane Cannon Foundry

13. Galata Tower
14. Sublime Porte
15. Covered Bazaar

Mosques and Mosque Complexes
16. Aya Sofya
17. Ahmed I (Blue Mosque)
18. Firuz *ağa*
19. Sokollu Mehmed *paşa*
20. Atik Ali *paşa*
21. Kara Mustafa *paşa*
22. Bayezid II
23. Yeni Valide
24. Rüstem *paşa*
25. Süleymaniye

26. Şehzade
27. *Fatih* Mehmed II
28. Selim I
29. Mihrimah
30. Kara Ahmed *paşa*
31. Ibrahim *paşa*
32. Ramazan *efendi*
33. Isa Kapı
34. Davut *paşa*
35. Haseki Hürrem
36. Azap Kapı
37. Mahmud *paşa*

Church
38. Greek Orthodox Patriarchate

Map 26: Apex and Decline of Poland, 16th–17th Centuries

Following the dynastic union of Poland and Lithuania effected by the marriage of Jadwiga and Jagiełło, Polish aristocratic culture took root among the Lithuanian nobility until, with the official treaty of union signed in Lublin (1565), the Lithuanians accepted direct Polish control over all their affairs, and Catholic Poland emerged as the dominant partner. But the successful dominance of the Poles over the Lithuanians exacted a heavy price—the Poles were forced to concentrate on the inherited problems posed by the centuries-old Lithuanian rivalry with Muscovite Russia, which lay in the east, to the detriment of their own essential interests, which historically lay in the west.

The Jagiełłonian dynasty died out in 1572, having presided over an internal fragmentation of political authority that left Poland-Lithuania one of the most decentralized and fragile states in Europe. At the expense of the royal throne, it had failed utterly to stem the tide of rising aristocratic political privilege. In 1505 the national *sejm,* the general assembly of the nobility, won royal recognition as the supreme decision-making organ in the state. This proved to be a catastrophe that rewarded the selfish particularism of the local nobles at the expense of state unity, removed virtually all central political authority, polarized Polish society between aristocratic "haves" and commoner "have-nots," stifled governmental reform, fossilized a feudal military system, and eventually made Poland's continued existence contingent on the interests and rivalries of foreign neighboring states.

With the end of the Jagiełłonian dynasty, the Polish nobility elected a series of weak foreign and native rulers in an effort to ensure their dominant political position. Poland became embroiled in the Livonian War (1558-82), begun by Russian tsar Ivan IV the Dread (1547-84) in an effort to win Russia ice-free ports on the Baltic. After dragging on indecisively for years, drawing in Sweden as an additional anti-Russian protagonist, the war came to a costly but victorious close for Poland. Russia was exhausted militarily, and, with the death of Ivan, strong central authority in the state collapsed, initiating the "Time of Troubles" (1598-1613).

Russia's Time of Troubles was the period of Poland's apex. Zygmunt III Vasa (1587-1632) was elected Polish king in an attempt to cement a more permanent anti-Russian coalition in the Baltic. While the Swedish Vasas were Protestant, Zygmunt had been educated by Jesuits and remained staunchly Catholic. Under him, the Counter-Reformation made rapid headway within Poland. In 1596 the Union of Brest created a new form of Catholic Christianity in Poland-Lithuania drawn from among Belorussian and Ukrainian Orthodox believers and known as the Uniate church.

Poland then started actively intervening in the chaotic internal affairs of Russia in an attempt to achieve a cherished Catholic and Jesuit goal—the conquest of Orthodox Russia by Catholic Poland. With Zygmunt's support, Polish and Cossack troops managed to place a pretender, Dimitri (1605-6), on the Muscovite throne, but his pro-Catholic actions led to a Russian reaction that reinstalled a native Russian, who was backed by Swedish allies, as tsar. Zygmunt invaded Russia behind a new pretender, captured Smolensk, negotiated with disaffected Russian nobles for his son's accession to the throne, and managed to occupy Moscow in 1610. The Swedes took advantage of the continued turmoil by seizing Novgorod. Zygmunt spoiled his triumph by reneging on the negotiations that would have made his son, Władisław, tsar by demanding the crown for himself. This about-face resulted in a popular protonationalist uprising in Russia that swept the Poles out of Moscow (1612). With roving bands of Poles, Swedes, and Cossacks continuing to devastate the countryside, the Orthodox Russians then elected a native tsar, Mihail I Romanov (1613-45), who won general Russian and Cossack support. In 1617 Mihail made peace with Sweden, winning back Novgorod. The Poles, pushed back from Moscow but still holding much Russian territory along the old border, were permitted to retain their border conquests by a truce signed in 1618.

Soon thereafter, Poland was embroiled in a new war with Sweden. In 1632 the truce with Russia expired and the Russians unsuccessfully reopened hostilities, ending in an "eternal peace" with the Poles (1634). Under this accord, Władisław VII Vasa, Zygmunt's son, now Polish king (1632-48), renounced his long-standing claim to the Russian throne. Then the Polish-Russian conflict shifted to Ukraine. When in the 1630s the hard-pressed Orthodox church in Ukraine turned to the Zaporozhe Cossacks for help in fending off the Poles' constant unionist pressures, they rose in revolt (1648), led by their *hetman* Bogdan Khmelnitsky, but failed. In 1654 Khmelnitsky turned to Moscow for help, and at Pereiaslavl he and the Muscovites signed a pact that placed the eastern regions of Ukraine under Russian protection, confirmed Zaporozhe Cossack autonomy, and elicited the sworn allegiance of the Ukrainian Cossacks and peasants to the Russian tsar. The conflict that erupted after Pereiaslavl, known in Polish history as the "Deluge," saw Poland invaded by Swedes, Russians, Cossacks, and Turks and the near collapse of the state. When the crisis was at its darkest, a miraculous defense of Częstochowa Monastery against the Swedes (1657) permitted Poland to survive and eventually stabilize its situation. In 1667 the Poles and Russians brokered an armistice at Andrusovo, in which Ukraine was partitioned: Russia gained Smolensk and eastern regions of Ukraine on the left bank of the Dnieper River, including Kiev; Poland retained western Ukraine and some territories on the right bank of the Dnieper. Desultory fighting broke out again soon thereafter and dragged on for another nineteen years. The conflict finally ended in another "eternal peace" concluded in 1686, in which the terms of Andrusovo were confirmed.

SWEDEN

To
Sweden

Novgorod

LIVONIA

Riga

COURLAND

Moscow

HOLY
ROMAN
EMPIRE

Smolensk

Andrusovo

RUSSIA

Berlin

Gdańsk

Vilnius

Oder R.

POLAND

Dnieper R.

MAZOVIA

Vistula R.

Vistula R.

Bug R.

Warsaw

Brest

Prague

Częstochowa

Lublin

UKRAINE

Kiev

HABSBURG

Cracow

L'viv

Pereiaslavl

GALICIA

Danube R.

ROYAL

PODOLIA

Vienna

HUNGARY

MOLDAVIA

Dniester R.

Dnieper R.

EMPIRE

Buda

Pest

CRIMEAN TATARS

Zagreb

TRANSYLVANIA

Danube R.

OTTOMAN

EMPIRE

MILES

| 0 | 100 | 200 | 300 |

Border of Poland-Lithuania at Greatest Extent (1634-1635)

Eastward Expansion, 1610-1619

Lands lost in 1667/86

KILOMETERS

| 0 | 100 | 200 | 300 | 400 |

Lands lost to Ottoman Empire, 1672; Returned to Poland, 1699

Border of Holy Roman Empire in Habsburg Lands

Polish Occupations of Moscow, 1610 and 1617/8

Map 27: The Rise of the Habsburgs, 16th–17th Centuries

The Habsburgs were the last to emerge of a series of royal houses that vied for possession of the thrones governing the medieval states of Central-Eastern and Northeastern Europe during the 14th through early 16th centuries. The family started out within the Holy Roman Empire as obscure minor German aristocrats holding lands in the eastern regions of Switzerland. The Habsburg rise from obscurity began in 1273 when Count Rudolf of Habsburg was chosen Holy Roman Emperor (1273-91) by the German Electors of the empire precisely because of his very obscurity and perceived weakness as a prince. During Rudolf's reign, the Grand Principality of Austria was established as the new heartland of Habsburg possessions. The family thereafter continued to enlarge its lands and to maintain claims on the imperial throne in Germany primarily through a deft political policy of marriage alliances that transformed the Habsburgs into the wealthiest and most politically powerful ruling house in all of Europe by the 16th century.

Habsburg Holy Roman emperor Frederick III (1440-93) married his son and successor, Maximilian (emperor, 1493-1519), to a Burgundian duchess, who brought as her dowry the vast riches and lands of that French house. In turn, Maximilian married his son, Philip the Fair, to Joanna, daughter and heiress of the Spanish monarchs Ferdinand and Isabella (1474-1504), who by sponsoring the voyages of Columbus opened to Spain the riches of the New World. The fruits of discovery fell into the hands of the fortunate couple, who produced two sons, Charles and Ferdinand. Between them, the brothers ultimately came to share the largest and wealthiest political inheritance in Europe. Drawing on their vast resources, the Habsburgs were thus able to exert their will on all those living within the orbits of the Holy Roman Empire and of Spain. No other European royal or princely dynasty had the wherewithal to compete directly with the family on its own terms.

After years of constant pressure from attempting to wield effective authority over his far-flung territories, and weighed down by constant warfare with Protestants, England, France, and the Ottoman Empire, the elder brother, Emperor Charles V (1519-56), concluded that the Habsburg possessions were unmanageable by a single head of the family. He abdicated the German imperial and Spanish royal thrones and divided the family patrimony between eastern and western lands, giving his brother Ferdinand I (emperor, 1556-64) the former and his son Philip II (king, 1556-98) the latter. The eastern branch of the divided Habsburg patrimony retained the imperial title and the core possessions of Austria. It was this eastern branch that continued to carry on the Habsburg legacy in Eastern Europe.

By the time Ferdinand I was elected emperor, he had already started Habsburg expansion beyond the patrimonial Austrian lands. His wife, Anna, was the sister of Louis II Jagiełło, king of Bohemia and Hungary, who was killed in 1526 on the battlefield of Mohács fighting the Ottoman Turks led by Süleyman I. As the brother-in-law and only direct male heir of Louis, Ferdinand won election to both vacant crowns following the debacle. His new Czech aristocratic subjects, however, refused to recognize him as hereditary king, and it became a major goal of the Habsburgs to gain hereditary right to the Bohemian throne. That issue caused bitter feelings between the two sides that ultimately led to the outbreak of the Thirty Years War in Bohemia (1618), the Czechs' crushing defeat at the Battle of White Mountain (1620), and the near total collapse of Bohemia's independence (1627), after which Slavic Bohemia was subjected to intensive Germanization, especially among the aristocratic and urban classes. Bohemia became directly integrated into the family patrimony of the Habsburgs under the guise of a fictional political autonomy.

Ferdinand's Hungarian election was contested by a large segment of the Magyar nobility of Transylvania, who advanced one of their own, János Zápolya, as rival monarch with Ottoman Turkish backing. In the wars against the Turks and Transylvanians that followed, the Habsburgs secured and consolidated a narrow strip of northern and western Hungary, known as Royal Hungary. (See Map 24.) They waged intermittent warfare with the Turks over the central Hungarian lands and Transylvania. Meanwhile, the Transylvanian Magyar nobility, taking advantage of the autonomy granted them by vassal status to the Ottoman Turks, forged their state into a powerful anti-Habsburg force during the religious wars of the 16th and 17th centuries. Under the powerful Báthory princely house in the second half of the 16th century, Transylvania increased its presence in Eastern Europe.

The decline of the Ottoman Empire in the 17th century opened the door to the Habsburgs for making good on their claims to the central Hungarian lands and Transylvania. Emperor Leopold I (1658-1705), having settled pressing European matters that had diverted Habsburg attention from the Turkish-Hungarian problem in the past, at last found himself free to deal decisively with the Turks. Fending off an Ottoman siege of Vienna in 1683, Habsburg forces countered with an all-out offensive that expelled the Turks from central Hungary and delivered Transylvania into Leopold's hands. In 1687, Leopold compelled a grateful Royal Hungarian nobility to declare the Habsburgs hereditary rulers of Hungary. Transylvania was not united with Royal Hungary. Through a separate imperial diploma issued by Leopold in 1691, it was placed directly under the authority of the Habsburg ruler as an independent royal principality. With the Treaty of Sremski Karlovci (1699), which ended the war of Hungarian reconquest in the Habsburgs' favor, the Turks were confined mostly to regions south of the Danube.

HOLY

ROMAN
● Nürnburg

EMPIRE

Danube R.

Augsburg
●

SILESIA

Prague
●

K. OF BOHEMIA

Cracow
●

POLAND

L'viv
●

GALICIA

Košice
●

ROYAL HUNGARY

MOLDAVIA

Vienna
●

Bratislava
●

THE AUSTRIAS

SALZBURG

Buda
● ● Pest

STYRIA

HUNGARY

Cluj
●

TRANSYLVANIA

TYROL

CARINTHIA

Danube R.

Senta
●

CARNIOLA

Venice
●

VENICE

Zagreb
●

Mohács
●

BANAT

CROATIA

DALMATIA

To
Venice

HOLY
ROMAN
EMPIRE

PAPAL
STATES

NAPLES

Sremski
Karlovci
●

Belgrade
●

WALLACHIA

Sarajevo
●

Danube R.

Dubrovnik
●

OTTOMAN   EMPIRE

―――――――	Border of Habsburg Empire. 1699
▨	Holy Roman Empire
▨	Hapsburg Lands withinHoly Roman Empire
●●●●●●●●●●	Eastern Border of Royal Hungary, 1526
▬ ▬ ▬ ▬	Border of Holy Roman Empire in Hapsburg Lands

MILES

0   50   100   150   200

0   100   200   300

KILOMETERS

Map 28: Ottoman Decline, 17th–18th Centuries

With the death of Sultan Süleyman I in 1566 and the succession by his son, Selim II the Sot, the Ottoman Empire slipped into gradual, inexorable decline brought on by both internal and external developments.

A unique facet of Ottoman government in its ascendancy was that all of its bureaucratic-military offices were staffed by the household slaves of the sultan, who held over them the absolute power of life and death. This made it the most centralized and efficient government in Europe from the 14th through the mid-16th century. It should be borne in mind, however, that slavery in the Ottoman context little resembled that of the American South—there were no degrading aspects involved. The sultan's slaves possessed and controlled immense power, wealth, social position, and public honor. Every government and standing military office was filled strictly on the basis of individual merit, with no regard whatsoever given to birth status or social position. As long as the sultans were capable rulers and generals and able and willing to exert their absolute authority over their slave household, the empire thrived. When the system began to break down in the mid-16th century, with jealous Muslim-born subjects forcing or bribing their way into government and military offices, thus loosening the sultans' absolute authority on which the entire stability of the system was dependent, and escalating internal corruption, the Ottoman Empire slipped into irredeemable decline.

During the 17th century the empire experienced a string of inept sultans. The effects of this loss of authority at the center were magnified by disruptive external pressures. Western European technologies played havoc on tradition-bound Islamic state reality. Beginning in the 16th century, naval developments in the West ushered in the "Age of Discovery," which opened sea routes to the necessary and lucrative spice trade with Far East Asia, circumventing the Ottoman middlemen who formerly controlled such commerce. Gold and silver from the Americas flooded the eastern Mediterranean markets of the empire, causing rapid inflation, higher taxes, and an explosive rush for cash on all levels of Ottoman society. Moreover, in the 17th century Western gunpowder technologies transformed the weaponry and tactics of warfare—a development to which the Ottomans, holding fast to traditional military approaches, were slow to respond. The result was the end of Ottoman military dominance in Eastern Europe and the onset of mounting defeats in battles with their Christian enemies. Western European states, such as France and England, were able to force treaties—the nature of which was reflected in their being named "capitulations"—on the Turks that placed nearly all of the empire's trade relations and profits in their own hands. Militarily antiquated and economically strangled, the Ottoman Empire ceased to expand, resulting in gradual, but perpetual, contraction in Europe.

The decline began with the disastrous second Ottoman siege of Vienna in 1683, following which by 1699 the Habsburgs succeeded in pushing the Ottomans out of Hungary and south of the Danube, except in the region of Banat. (See Map 27.) The Turks were also plagued by a continuing war with Russia north of the Black Sea, which ended in 1702 and resulted in further territorial losses. These were regained in 1711 when Tsar Peter I the Great (1682-1725), posing as champion of the Ottomans' Balkan Christian subjects, unsuccessfully invaded Moldavia, found himself encircled by the Turks, and was forced to return past acquisitions to save his army and reputation. Ottoman success proved transitory. The Habsburgs reopened military operations along the Danube in 1716, defeated the Turks at Petrovaradin, and, led by Prince Eugene of Savoy, captured Belgrade (1717). In the Treaty of Požarevac (1718), the Ottomans lost Banat and relinquished areas of northern Serbia and western Wallachia (Oltenia) to the Habsburgs.

The Ottoman military, however, had not yet completely collapsed. This fact was demonstrated in a renewed war with both the Habsburgs and Russia that began in 1736. Though Russia succeeded in winning territories along the northern Black Sea, the Turks managed successful campaigns against both their enemies in 1737. By the terms of the Treaty of Belgrade (1739), the Habsburgs relinquished control of that city, northern Serbia, and Oltenia.

The War of the Austrian Succession (1740-48) and the Seven Years War (1756-63), in which the Ottomans' major European enemies were involved, granted the Turks a period of peace in the Balkans. But the respite ended in 1768 when Catherine II the Great (1762-96) sent Russian troops into Moldavia and Wallachia in pursuit of defeated Polish rebels who had fled to the Ottoman Empire. Sultan Mustafa III (1757-74) declared war but was unable to prevent the Russians from overrunning the two principalities or from conquering Crimea. Russian military successes in the war were partly responsible for Prussian king Frederick II the Great's (1740-86) arranging the first partition of Poland with Catherine in an effort to counterbalance her rising power in Eastern Europe. (See Map 29.) Only the need to concentrate military resources to crush the great Cossack revolt led by Pugachev (1773) diverted the Russian empress from her goal of conquering Istanbul and resurrecting an Orthodox imperium in the Balkans.

The Treaty of Kyuchuk Kainardzha (1774) reflected the Ottomans' diminished status as a European Great Power. Russia gained extensive territories in the Black Sea region and free commercial navigation of its waters. A controversial clause also gave Russia representation at the Porte on behalf of the Turks' Orthodox subjects, while the Turks promised to protect Orthodox church property throughout the empire.

HOLY
ROMAN
EMPIRE

POLAND

RUSSIA
(after 1772)

Danube R.

Vienna

BUKOVINA

HABSBURG EMPIRE

Buda • Pest

MOLDAVIA

CRIMEA

VENICE

Zagreb

TRANSYLVANIA

Venice

CROATIA

Petrovaradin

BANAT

WALLACHIA

PAPAL
STATES

DALMATIA

To
Venice

BOSNIA

Sarajevo

Belgrade

Požarevac

OLTENIA

Bucharest

SERBIA

Vidin

Danube R.

Niš

Kyuchuk
Kainardzha

Dubrovnik

MONTENEGRO

Sofia

Naples

OTTOMAN EMPIRE

NAPLES

ALBANIA

Thessaloniki

Istanbul

Ioannina

To
Venice

Athens

K. of
SICILY

MILES

| 0 | 50 | 100 | 150 | 200 |

| 0 | 100 | 200 | 300 |

KILOMETERS

——— Boundary of the Ottoman Empire in 1774

Lands temporarily lost to Habsburgs, 1717-1739

Lands permanently lost in 1718

Lands permanently lost in 1774

Map 29: The Partitions of Poland, 1772–1795

The most critical factor in Poland's fall from greatness was the state's lack of a viable, unifying sociopolitical system to hold the country together. By the mid-17th century, Poland was disintegrating from within. The Polish nobility had grown so powerful vis-à-vis royal authority that no real unity of purpose existed among them to defend the interests of the state as a whole. Government through the *sejm* was paralyzed by the use of the *liberum veto,* which granted each individual member the right to defeat any resolution placed before it by voicing a lone protest. No governing assembly could function under such a condition. Although the vetoes of lesser nobles could be overcome effectively, those of the powerful magnates, as well as those with strong foreign backing, could not.

Frustrated by the situation, in 1668 the Vasas relinquished the Polish throne. A long series of succession struggles ensued, extending into the 18th century. The Polish throne became a pawn for furthering the interests of various European Great Powers in the balance of power game they played among themselves. By manipulating the *liberum veto,* Russia, Prussia, the Habsburgs, and France sought to better their positions vis-à-vis each other through installing puppet Polish kings and keeping the Polish nobility divided and weak. Oftentimes foolhardy policies undertaken by the unfortunate Polish rulers magnified Poland's problems. For example, King Augustus II the Strong (1697-1733) allied himself with Russia to despoil Sweden of its Baltic possessions. The Great Northern War (1700-21) resulted. Fought mostly on Polish territory, it led to widespread devastation inside Poland, Augustus's overthrow and subsequent forced return, and further internal political deterioration.

The election of Augustus's successor opened the door to wholesale foreign machinations inside Poland. The Poles elected Stanisław Leszczyński with French support, since their candidate was the father-in-law of French king Louis XV (1715-74). Russia considered Leszczyński unacceptable because of his pro-Swedish sympathies and garnered support from the Habsburgs, who, in turn, insisted on the election of Augustus's son. The Russians invaded Poland and drove out Leszczyński, sparking the War of the Polish Succession (1733-35) among the Great Powers. The war ended in 1735 with Russian-Habsburg policy in Poland victorious.

King Augustus III of Saxony (1734-63), the successful Russian-Habsburg candidate, expended little interest or time on his kingdom. During the two decades of his disinterested reign, Russia increased its encroachment into Polish internal affairs, causing a growing amount of anti-Russian sentiment to arise among a segment of the nobility led by the Potocki family. It was opposed by another aristocratic faction led by the Czartoryski family, which proposed strengthening royal authority, abolishing the *liberum veto,* and other more centralizing political reforms. The Czartoryski group looked to Russia for backing.

In 1764 the Czartoryskis, supported by Catherine II the Great, who considered Poniatowski a malleable puppet because of a prior intimate relationship, elected Stanisław II Poniatowski (1764-95) king. In the process, Russia and Prussia formulated a cooperative policy aimed at keeping the Polish state weak and docile to their increasingly common imperial interests. Once king, however, Poniatowski attempted to act independently of his powerful Russian patroness and to institute the Czartoryski reform program. Russia invaded Poland and Prussian king Frederick II the Great, fearful of renewed military involvement in a costly conflict, struck upon the idea of partitioning segments of Poland among the powers directly interested in Polish matters. In the deal forged among Russia, Prussia, and the Habsburg Empire in 1772, Russia was granted Belorussia and slices of Polish Ukraine; the Habsburgs received Galicia; and Prussia stole West Prussia, less the city of Danzig (Gdańsk). The First Polish Partition stripped Poland of a third of its territory and almost half of its inhabitants.

Poland was left no choice but to accept the partition. Too late to save their state, the Poles instituted political reforms. A new constitution was promulgated (1791) that transformed the elective kingship into a hereditary one, vested executive power in the royal office, placed legislative power in the hands of the *sejm,* and abolished the *liberum veto.* Prussia and the Habsburgs accepted the reforms. Russia opposed them and invaded Poland, which sparked a similar act by the Prussians, who feared being left out of any possible territorial aggrandizement that might result. To avoid further bloodshed, another bargain was struck between the two invading powers at Poland's expense. In 1793 the Second Polish Partition was implemented. Russia took most of both historic Lithuania and Polish Ukraine. Prussia pilfered Danzig and Great Poland. Moreover, the Poles were forced to accept an alliance with Russia, granting the Russian military the right of free entry into what was left of Poland, as well as Russian control of rump Poland's foreign relations.

By the time of the two partitions, the Western concept of nationalism had sunk firm roots among the Polish nobility. Within a year of the second partition, a national uprising erupted. Vastly outnumbered by the forces of both Russia and Prussia, the rebels, led by Tadeusz Kościuszko, put up a brief heroic but futile fight. Kościuszko was captured and Warsaw was taken by the Russians. In 1795 the Third Polish Partition was implemented. By its terms, Russia grabbed what remained of Lithuania and Ukraine, as well as Courland; Prussia took Mazovia with Warsaw; and Habsburg Austria, not to be left out of the final division of Polish spoils as it had been in 1793, obtained what remained of Little Poland. Poland ceased to exist as a state.

SWEDEN

COURLAND

Riga

POLISH
LIVONIA

SAMOGITIA

Polatsk

Vitsebsk

Kaunas

Königsberg

Vilnius

Dnieper R.

PRUSSIA

PRUSSIA

Danzig

WEST
PRUSSIA

LITHUANIA

Smolensk

RUSSIA

MAZOVIA

BELORUSSIA

Minsk

Poznań

Vistula R.

GREAT
POLAND

Bug R.

Warsaw

SILESIA

Breslau

LITTLE
POLAND

Pripet R.

Cracow

GALICIA

L'viv

VOLHYNIA

UKRAINE

Kiev

Košice

Dniester R.

PODOLIA

HABSBURG
EMPIRE

Vienna

Danube R.

Buda

Pest

OTTOMAN
EMPIRE

——— Border of Poland in 1771
——— Border of Poland following the First Partition, 1772
- - - - - Border of Poland following the Second Partition, 1793
– – – Central European Borders after the Third Partition 1795
▥ Polish Lands taken by Russia
▨ Polish Lands taken by Prussia
▤ Polish Lands taken by the Habsburg Empire

MILES

0   50   100   150   200

0   100   200   300

KILOMETERS

Part IV

Period of Nationalism

(19th Century–1918)

Map 30: Eastern Europe, 1812

When the final two partitions of the Polish state were implemented, the French Revolutionary Wars, which started a process that forever changed the core political culture of Western Europe, had already begun. Along with "Liberty, Equality, and Brotherhood," the armies of the revolution spread the notion that the borders of a liberal-democratic "nation" were somehow sacred. As the Poles dropped into "national" oblivion, the idea of an innate, "enlightened" sacredness about nations and their borders was commencing its successful rise to prominence in Western European political consciousness.

The humiliated and discontented Polish nobility were a receptive audience for the nationalist revolutionary message of France. After French emperor Napoleon I Bonaparte (1804-15) transformed the ideology of the revolution from radical republicanism into nationalist-imperialism, the Poles came into direct contact with the French during Napoleon's 1806-7 Jena-Auerstädt, Eylau, and Friedland campaigns against Prussia and Russia. Playing to the Poles' agitated national awareness, Napoleon issued calls for uprisings in the Polish territories held by his two enemies. Following Prussia's military demise at Jena-Auerstädt (1806), Napoleon advanced into former Polish lands against the Russians and established his headquarters at Warsaw. The Polish nobility grew nationalistically ecstatic. Napoleon's appearance removed Russian and Prussian administration from much of the formerly Polish territory, and Napoleon himself seemed to hold out the promise of imposing a resurrected Greater Polish state on the now defeated partitioning powers.

With Napoleon's defeat of the Russians at Friedland (1807), the last major military obstacles to French dominance in Central-Eastern and Northeastern Europe were shattered, and both Russia and Prussia were forced to sign the Treaty of Tilsit. Prussia ceded to Napoleon all lands taken from Poland since the first partition for the purpose of creating the Grand Duchy of Warsaw (the name "Poland" was not used, at the insistence of Russian tsar Alexander I [1801-25]). In turn, Russia was constrained to officially recognize the independence of the new "Warsaw" state. Far from satisfying the national ambitions of the Polish nobility for a truly independent Poland, the Grand Duchy was placed under direct French suzerainty. Napoleon looked on it as a useful tool for furthering French imperial interests—the vengeful and nationalistic Poles would serve as a Damocles' sword for keeping the defeated states in line with French dictates. He gave only ambiguous support to the Poles' nationalist aspirations—just enough to keep them expectantly loyal but not enough to give them real freedom of action.

Because of the illusion, created by Napoleon's political savvy, that Polish national resurrection lay just after the next successful military campaign, Poles fought loyally and bravely as France's dependent and dependable allies until the bitter end arrived for Napoleon at Waterloo in 1815. An army-size Polish military force accompanied Napoleon in his ill-fated 1812 invasion of Russia, the Poles' most intransigent enemy, and additional Polish troops were among the most trustworthy units in his desperate attempts to stave off total military defeat by the combined, overwhelming forces of all the European Great Powers during 1813 and 1814. Poles formed part of Napoleon's private personal guard during his exile on Elba, and Poles died in the carnage at Waterloo attempting to return their idol, and potential national benefactor, to imperial power in Europe.

Napoleon's military conquests also had a decisive impact on developments in the Germanic regions of Europe, and on the Habsburg Empire especially. His defeat of a combined Prussian, Habsburg, and Russian force at Austerlitz (1805) resulted in the termination of the Holy Roman Empire in 1806. The Habsburgs were forced to renounce that throne, retaining only the imperial office for lands constituting the family's personal patrimony—the Austrias, Bohemia, Hungary, Transylvania, and Galicia—thereafter known collectively as the Austrian Empire. The former German states of the disbanded empire were consolidated and organized as the pro-French Confederation of the Rhine; Habsburg Venetian territories, won in 1797, were handed to the French puppet Kingdom of Italy. After the Austrian Empire suffered another defeat by the French in 1809 at Wagram, most of former Hungarian Croatia and Habsburg Slovenia were transformed into the Illyrian Provinces and directly incorporated into Napoleon's French Empire (1809-13).

The incorporation of much of the Croats' and Slovenes' territory into Napoleonic France opened the door to their direct exposure to liberal and national concepts. Under this new arrangement, French administration, French law, and French language were imported into the provinces. The French instituted public works projects (road and bridge building, reforestation, land reclamation) and social reforms (serf emancipation, land redistribution). Most important, direct French rule not only brought the ideas of liberal democracy, nationalism, and the nation-state to the provinces, it put them into actual practice. Although the French interlude lasted only a few short years, and its thrust was essentially anti-Croat and anti-Slovene (since the provinces were considered *French* and not Croatian or Slovenian), the Croats and Slovenes gained a heightened awareness of their own ethnonational identities and briefly experienced firsthand the benefits of the new liberal, nationalist political culture that France heralded for the other member societies of the West.

In the Ottoman Balkans, nationalist concepts spread from Serb emigrants in Habsburg Slavonia to the Serb Ottoman subjects, transforming a local uprising begun in 1804 in the Belgrade area into a nationalist Serb movement by 1815. (See Map 32.)

PRUSSIA

Tilsit
Vilnius

Eylau Friedland

Minsk

Hanover Berlin

DUCHY
OF
WARSAW

Warsaw

RUSSIA

CONFEDERATION

Auerstädt Leipzig

Jena

SILESIA

Prague

BOHEMIA

OF THE

Cracow

GALICIA

L'viv

Kicv

Austerlitz

AUSTRIAN

Munich

Vienna Wagram

RHINE

Danube R. Bratislava

HUNGARY

BESSARABIA

K. OF
ITALY

Ljubljana

Venice

ILLYRIAN
PROVINCES

Zagreb

EMPIRE

Buda Pest

Danube R.

Cluj

TRANSYLVANIA

MOLDAVIA

Iaşi

SLAVONIA

Sarajevo

PAPAL
STATES

Zadar

BOSNIA

Belgrade

WALLACHIA

Rome

SERBIA

Bucharest

Danube R.

Dubrovnik

MONTENEGRO

Sofia

K. OF
NAPLES

Naples

OTTOMAN
EMPIRE

Thessaloniki

Istanbul

CORFU

SICILY

MILES
0 50 100 150 200

0 100 200 300
KILOMETERS

Athens

French Imperial Possessions

French Dependent Allies

Map 31: Eastern Europe, 1815

Napoleon's disastrous Russian Campaign (1812) led to the collapse of the French Empire within two years under the combined weight of forces from Russia, Prussia, and Habsburg Austria, supported by England. Following an invasion of France by his enemies and his abdication in 1814, the victorious powers gathered in Vienna to reshape the post-Napoleonic European world. Eastern Europe played an important role in the decisions of the victors.

At the Congress of Vienna in 1815, the victorious allies attempted to redraw the map of Europe for their own benefit and to suppress any future threat to the old monarchical order that could be posed by liberal democracy and nationalism by imposing a police-state regime in all territories under their control. The Austrian Habsburg and Prussian Hohenzollern monarchies were restored to complete independence. The Habsburgs received all territories they formerly held prior to Napoleon, as well as new Italian possessions, including Venetia. The Illyrian Provinces were reclaimed, and Bavaria handed over lands in its south. Prussia received western slices of the Duchy of Warsaw and Danzig. To replace Napoleon's Confederation of the Rhine (and, indirectly, the Holy Roman Empire), a defensive alliance of thirty-nine small German states, known as the German Confederation, was created. (It was finalized only in 1820.)

As a part of the remapping scheme, most of the Grand Duchy of Warsaw, composed of partition spoils originally held by Prussia and partly by Austria, was handed over to Russia. Tsar Alexander insisted on Russia's being compensated for its efforts in defeating Napoleon. Since Russia had played a leading role in the success and possessed the single largest military force in Europe at the time, the rest of the allies were obliged to appease him with the Polish spoils he sought, especially as they desired similar compensation for themselves in other areas. With the territorial reward attained at Vienna, Russia came to hold the largest and most populous segment of former Poland.

Alexander organized his new Polish acquisitions into an autonomous Kingdom of Poland (commonly called "Congress Poland") in permanent union with Russia, with the Russian tsar serving as hereditary ruler. The Poles were granted an administrative system separate from that of Russia, a national *sejm,* and their own military forces. They were permitted to continue using their native language in all official capacities. A viceroy represented the tsar inside the kingdom and Russian generals were placed in overall command of the Poles' military. On the whole, the position of the Catholic Poles within the Orthodox Russian imperial state was not overly negative. A fiction of Polish national existence was maintained by the Russian-imposed royal constitution, and the Polish aristocracy were confirmed in their dominance within the kingdom. Adam Czartoryski, a Polish prince, was a longtime friend of Tsar Alexander who had served as an important Russian minister during the wars and had helped influence the tsar in his creation of the kingdom. For a brief period of time after the establishment of the kingdom, Czartoryski and many of his fellow Polish nobles thought that Alexander would even further the Poles' nationalist cause by turning over control of vast Ukrainian territories to them. But they were proved mistaken. After 1820 Alexander gradually succumbed to a personal inclination toward Orthodox religious mysticism, and he became unwilling to abide by the terms of the Polish kingdom's constitution. The Polish nationalist aristocracy thereafter increased their agitation for immediate satisfaction of an extreme nationalist agenda, resulting in the failed 1830 Polish Revolution.

To compensate for the decline in perceived political prestige suffered during the wars (caused by the loss of the Holy Roman Empire and less than admirable military efforts), as well as to bend all of their remaining actual European political influence toward preserving their existence, following the congress the Habsburgs led the reactionary efforts of the victorious European monarchies to suppress liberal and national movements in Europe. It was no accident that the last but futile effort to stem the tide of liberal democracy and romantic nationalism was born in Habsburg Vienna and became associated with the name of the Austrian chancellor Klemens von Metternich (1809-48). A creature of an outdated political system, his only answer to the powerful liberal and nationalistic ideologies of the French Revolution was to turn back the political clock by autocratic force. He convinced his allies that monarchism could be defended from liberal democracy through autocratic authority exerted by means of repressive legislation, increased police enforcement, and strict media censorship. If these should fail to stamp out liberal threats in any of the monarchical states, then the other allies could be called on to intervene militarily. They could coordinate their policies and mediate any disputes that might arise through periodic meetings (congresses) of representatives at the highest governmental levels.

This swan song of the old political order in Europe lasted thirty-three years and never functioned entirely as planned. The congress idea lasted little over a decade before it collapsed after both England and France opted out. More significantly, judicial repression, police surveillance, and censorship (then as now) could not destroy or effectively hinder the spread of ideas considered dangerous to the state, ideas that the victorious monarchies themselves had fostered in the first place to raise the mass armies that eventually brought them victory in the wars against Napoleon.

The future portent of those ideas for the Vienna allies was then being demonstrated by Serbs and Greeks in the Balkans, whose national revolutions were breaking up the old order of the Ottoman Empire. (See Map 32.)

In 1804 a Serb uprising inside the empire erupted in reaction against the arbitrary rule of the Turkish governor of Belgrade. The rebels were aided by the Ottoman central government, which armed them to assist in bringing the Belgrade anarchists under government control. The Serb rebels were led by Djordje Petrović, a prosperous pig dealer whose large size and swarthy complexion earned him the nickname of Karadjordje (Black Djordje), and whose profession and former experience as a sometime bandit won him support among the Serbian middle and renegade classes.

The Serb rebellion attracted much support among Serbs whose ancestors in the late 17th and throughout the 18th centuries had fled the Ottoman Empire to settle in the Habsburg-held regions of Vojvodina and Slavonia, where they enjoyed certain cultural privileges granted them in return for services along the military border with the Turks. They were exposed to Western and Russian intellectual currents, including the new concepts of secularism and nationalism. When the Serb uprising erupted, they sent volunteers and supplies to Karadjordje. The emigré Serb rebels also attempted to establish a military alliance with Russia, which went to war with the Turks in 1806, the same year Karadjordje's rebels captured Belgrade. Soon the influence of the emigré Serbs transformed the rebellion from an attempt to reestablish legitimate Ottoman rule into a struggle for Serb independence. When the Russians signed an armistice with the Turks in 1807, however, the rebels were left facing the full brunt of available Turkish forces. Reprieved by renewal of the Russo-Turkish war in 1809, the Serbs found themselves abandoned when Russia was compelled to sign the Treaty of Bucharest (1812) so it could face Napoleon's impending invasion. The rebels then fell into internal disarray and the Turks reoccupied their territories in 1813. Karadjordje fled to the Austrian Empire, and the uprising was temporarily stifled.

In 1815 the Serbs again rose against the Turks. This time they were committed to establishing their complete independence from Ottoman control. Karadjordje was still in Western exile, so his place as rebel leader was taken by Miloš Obrenović, a middle-class individual with an abiding personal hatred of Karadjordje, whom he accused of having poisoned his half brother. Through dogged military leadership, astute diplomacy, and highly refined bribery of Turkish officials, Obrenović won recognition from the Turkish government as prince of an autonomous Ottoman province of Serbia in 1817. When Karadjordje returned and voiced opposition to Obrenović's seemingly pro-Turkish approach, he was assassinated by Obrenović's supporters, sparking a deadly blood feud between the two future Serbian royal families that would plague Serbian politics into the 20th century.

By virtue of their privileged commercial position in the Ottoman Empire, the Phanariote Greeks maintained direct relations with Western Europe and Russia. Greek trading colonies were found throughout Europe and the Russian Black Sea coast and served as channels through which Western and Russian ideas spread to the Greek merchant class. The Greek colonists evolved into a Greek nationalist vanguard. With the founding of the Society of Friends (*Philike hetairia*) in 1814, secret revolutionary societies sprang up in numerous Greek merchant colonies throughout Europe. They turned to Orthodox Russia for support of their plans, especially when one of their number, John Capodistrias, became Tsar Alexander I's foreign minister.

In 1821 a Greek nationalist revolution against Ottoman rule erupted when Greek merchants' agitation for action against the Turks coincided with Russian imperialist aims at dominating the Balkans and opening the Mediterranean to Russian naval activity. A Russian force, led by a Greek general, Alexander Ypsilantis, unsuccessfully attempted to take control of the Romanian Principalities, which the Greek nationalists mistakenly considered Hellenized by a century of Phanariote rule. Though a failure, news of the episode sparked a rebellion among the oppressed Greeks in Greece proper, who conducted an initially successful guerilla war against divided and weak Ottoman forces in the region.

The Turks' first reaction was to hang the Greek patriarch from the gate to his cathedral in the Phanar. Then, in 1824, they called in Egyptian troops, the only effective military forces left in the empire, to quell the uprising. In the midst of the fighting, the rebels began squabbling among themselves. As Egyptian successes mounted, Western Europeans, imbued with philhellenic sympathies and by then in the throes of the Romantic Movement, grew fed up with the slaughter of fellow Christians in Greece by Muslims. Western volunteers (including the poet Byron, who died at Mesalóngion) streamed into the region to fight on the side of the rebels, whom they mistakenly considered the direct descendants of the classical ancients. Public opinion in England, France, and Russia eventually overcame the official concept of nonintervention in revolutionary wars implicit in the policies established at Vienna. Those states dispatched fleets to the eastern Mediterranean, and in 1827 their combined intervention brought the destruction of the Egyptian-Ottoman fleet in Navarino Bay, effectively resulting in the Greeks' complete independence from the Turks in 1829.

By the Treaty of Edirne (Adrianople) in 1829, which ended yet another Russo-Turkish war (1828-29), Serbia was recognized as an autonomous state within the Ottoman Empire, governed by a hereditary princely family, the Obrenovićes. The Turks also recognized the independence of a small state of Greece, whose capital was Navplion (later, Athens), governed by a king, Otto I (1832-62) from Bavaria. Wallachia and Moldavia were essentially occupied by Russia and direct Turkish or Greek influence was permanently removed.

BAVARIA

Danube R.

Vienna

RUSSIA

AUSTRIAN EMPIRE

Buda • Pest

HUNGARY

Iaşi

BESSARABIA

Zagreb

CROATIA - SLAVONIA

Venice

Cluj

TRANSYLVANIA

MOLDAVIA

Novi Sad

VOJVODINA

Belgrade

BOSNIA

Zadar

DALMATIA

Sarajevo

SERBIA

WALLACHIA

PAPAL
STATES

Mostar

Bucharest

Vidin

Ruse

Danube R.

Dubrovnik

MONTENEGRO

Niš

BULGARIA

Sofia

Skopje

Naples

Plovdiv

Edirne

Bitola

Thessaloniki

Istanbul

KINGDOM

OF

THE

TWO SICILIES

Ioannina

Mesalóngion

Athens

GREECE

*CYCLADES
ISLANDS*

Navplion

Navarino

MILES

| 0 | 50 | 100 | 150 | 200 |

| 0 | 100 | 200 | 300 |

KILOMETERS

To
Egypt

CRETE

OTTOMAN EMPIRE

Ottoman Empire
Autonomous Serb and Romanian States
Independent Greece

Map 33: Revolutions in the Austrian Empire, 1848–1849

The 1848-49 revolutions in the Habsburg Austrian Empire originated in the reign of Emperor Joseph II (1780-90). Joseph attempted to create a unified regional state similar to other contemporary European monarchies by instituting reforms aimed at centralizing the empire's government in Vienna. Among Joseph's initiatives was his attempt to make German the common administrative language. His reforms provoked a reaction among the empire's non-German populations, whose aristocrats and intellectuals turned to emphasizing local rights, traditions, and cultures. Upon Joseph's death, most of his reforms were revoked. To mollify rising fears concerning Germanization, the Habsburgs then established a chair of Slavic languages at the University of Vienna (1791), giving semiofficial sanction to the cultural validity of the empire's non-German subjects.

The Hungarians were the first of those peoples to take advantage of the situation. Since the 11th century the literary language of the Magyar nobility had been Latin, no matter the ethnicity of the individual. Beginning in the 1770s a movement to replace Latin with Magyar began, reinforced by reaction against Joseph II's Germanization efforts. By the 1840s the native language movement turned liberal and nationalist but split between two factions. The moderates, led by István Széchenyi, called for the overthrow of traditional aristocratic leadership in favor of liberal institutions and partnership with non-Magyars in a multicultural Hungary within a reformed Habsburg Empire. Radicals, led by Lajos Kossuth, demanded immediate liberal reforms favoring the traditional aristocracy and complete independence for the Kingdom of Hungary as an exclusively Magyar nation-state.

The Czech answer to Habsburg initiatives was Panslavism, born in Prague and nurtured by the creation of the Slavic language chair at the University of Vienna. Started in 1792 by Josef Dombrovský, Panslavism was a scholarly, intellectual, and romantic movement among Slavic philologists, who traced their linguistic history back to the time before the Slavs divided into the three main language groups. They stressed the common origin of all Slavic languages and the supposed brotherhood of all Slavic peoples, and looked toward a future when all the Slavs would be united equally in a Great Slavic confederation, which would naturally be shaped around Russia, the only independent Slavic state then existing. The Panslav Czech intellectuals in Prague soon attracted the attention of linguistically related Slovak intellectuals (such as Pavel Šafařík and Jan Kollar), who moved to Prague and added their voices to the rising Panslav chorus.

Among the Croats, the Napoleonic Illyrian Provinces episode opened the door to direct exposure to liberal and national concepts in their most militant form. (See Map 30.) They possessed a strong aristocracy that had been politically active in the Hungarian diet and provincial administration prior to the coming of the French, and their native literary language could be traced back to the 13th century. By the 1840s the South Slav movement, often called the Illyrian Movement, emerged among the aristocracy, led by Ljudevit Gaj. Advancing the myth that the ancient Illyrians were actually Slavs, they held that, because of their history, culture, and political capabilities, they were superior to all other Slavic peoples in the Balkans and rightfully deserved to lead any future Balkan Slavic confederation that might emerge.

When the February 1848 Revolution erupted in France, it sparked uprisings in various German states and in the Habsburg Empire calling for the establishment of constitutional governments. Pressurized by the continental monarchies' post-Vienna reactionary repression, radical liberalism swept through Central and Central-Eastern Europe and aimed at overthrowing the political order personified by Metternich. The March riots that erupted in the streets of Vienna, the nationalist revolutionary assemblies that convened in Budapest and Prague, and the Panslav Congress staged by the Czechs in Prague struck a telling first blow to the monarchies of Germany and Austria, which initially were shaken enough to grant liberal concessions to the various assemblies. Metternich resigned and fled the Habsburg Empire.

The tide began to turn against the revolutionaries in June when the Habsburg military crushed the Czech nationalists and Panslavs in Prague. In that same month the Croats declared their independence from Hungary and Baron Josip Jelačić led a Croat invasion of Hungary with Habsburg blessings. The Hungarians responded by intensifying their demands for self-rule and by unsuccessfully invading Austria proper in October. Vienna, again wracked by revolutionary agitation, was bombarded into submission to Habsburg authority, and its radical revolutionary leadership was executed.

Habsburg emperor Ferdinand I (1835-48) abdicated in December. His successor, Francis Joseph (1848-1916), considering himself free from any constitutional promises his predecessor had made, led the counterattack against the Hungarians, the last remaining revolutionary force in the empire. A constitution ensuring the centralized autocratic authority of the Habsburg emperor, despite the facade of a representative state diet and a responsible ministry, was promulgated in March 1849 for the empire as a whole. The Hungarians then proclaimed independence from Habsburg rule and declared Hungary a Magyar nation-state. Kossuth was elected governor-president by the diet and laws discriminating against non-Magyars were promulgated. Wracked by Serb and Romanian revolutionary movements in Banat and Transylvania, and by Austrian and Croatian invasion, independent Hungary collapsed in August when Tsar Nicholas I (1825-55) sent Russian troops against the Hungarians. By year's end, the Hungarian forces were crushed and Hungary was firmly back in Habsburg hands. The Habsburg monarchy had weathered the revolutionary storm.

PRUSSIA

RUSSIA

Sadova

Prague

SILESIA

Cracow

K. OF BOHEMIA

GALICIA

BAVARIA

Danube R.

Munich

K. OF
HUNGARY

BUKOVINA

Vienna

Bratislava

AUSTRIAN

EMPIRE

Buda

Pest

Pákozd

Bistriţa

TRANSYLVANIA

Cluj

Sighişoara

Danube R.

Sibiu

VENETIA

Ljubljana

Venice

Zagreb

Világos

Timişoara

CROATIA-SLAVONIA

BANAT

WALLACHIA

Zadar

BOSNIA

Belgrade

TUSCANY

DALMATIA

Sarajevo

SERBIA

Danube R.

PAPAL
STATES

Split

OTTOMAN EMPIRE

Dubrovnik

MONTENEGRO

—————— Border of Austrian Empire
- - - - - Border between Austria and Kingdom of Hungary

▬▬▬► Operations of Jelačić, 1848

▬▬▬► Main Austrian Military Operations, 1848-1849

▭▭▭▷ Russian Interventionist Forces, 1849

MILES

0   50   100 150 200

0   100 200 300

KILOMETERS

Map 34: The Austro–Hungarian Ausgleich, 1867

The rise of European nationalism, and its corollary the nation-state, spelled doom for the essentially medieval Habsburg Austrian Empire, which had managed to survive the convulsions of 1848-49 by the barest of margins. This success stemmed from the Habsburgs' ability to play off one nationalist group against another. The diverse ethnonational makeup of the empire's population made anti-Habsburg unity difficult to achieve. Finally, it was German nationalism, and not the national aspirations of the Habsburgs' non-German subjects, that actually undermined the continued existence of the medieval state that they controlled.

The 1848 German Revolution was a failed attempt of allied liberals and nationalists to create a single, constitutionally united Germany out of the German Confederation. After their initial successes, the revolutionaries spent too much time debating the issue of a united Germany's nature, thus the monarchical authorities were able to regain control and the revolution ended with the forced reversion to the status quo ante. If a German nation-state was to be constructed, it would be through the will of its monarchs.

Among Germans, the practical nationalist result of 1848 was the development of two separate concepts regarding the nature of a future German nation-state. There was the traditional, medieval idea of Greater Germany, which would have included all territories historically linked to German identity, including the Catholic South German lands of the old Habsburg-led Holy Roman Empire. On the other hand, there now emerged the Lesser Germany ideal, which accepted the loss of certain traditional German lands—those in the Habsburg Empire—to build a strong German nation-state from among the areas that were closely related culturally through Protestantism and almost exclusively German ethnically. The Habsburgs never were faced directly with the need to make a decision—it was made for them by Otto von Bismarck, chancellor of Prussia (1862-90), in 1866, when Prussia defeated Austria in the Battle of Sadova (Königgrätz). In the peace treaty that followed, the Habsburgs were forced to give up their Italian possessions and accept exclusion from further direct involvement in the affairs of northern Germany, which in 1867 Bismarck unified into a new constitutional confederation of German states under Prussian leadership.

Humiliated in war and unable to cow rabid Hungarian nationalists who constantly and actively resisted the harsh treatment meted out to them by imperial authorities following 1849, the Habsburgs were forced to come to grips at last with the realities of their medievalism. They worked out a political compromise with the only consistently powerful nationalist group in the empire—the Hungarians—attempting to preserve some semblance of real state strength in European Great Power politics by making their most dangerous internal threat a governing partner. For all practical purposes, the Compromise (Ausgleich) of 1867 created virtu-

ally two separate states from the Austrian Empire—Austria, including Bohemia and Slovenia, which was governed directly by the Habsburgs as patrimonial possessions, and Hungary, comprising the historic lands of the medieval Crown of St. István (Hungary Proper, Slovakia, Transylvania, Croatia, and Banat) claimed by the Magyar nationalists, which was governed by the Hungarian diet meeting in Budapest. Both halves were held together by the common rule of the Habsburg monarch—emperor of the family patrimony and king of Bohemia in the Austrian half, king alone in the Hungarian. Further bonds were provided by the imperial government's retention of control over crucial military, foreign policy, and financial matters common for both states. In effect, the Ausgleich legally recognized the Habsburgs' inability to continue ruling a multinational state in a 19th-century European world—that was being rapidly transformed by nationalist nation-state political reality—in an arbitrary 16th-century imperial fashion.

Through the Ausgleich, the monarchy managed to preserve both the borders of its ancient state and an outward semblance of continued political strength by conciliating and supporting the most cantankerous of its increasingly fractious harem of peoples. But there was no disguising the internal political weaknesses of the empire in its new form. The compromise could not help but agitate the less fortunate non-German and non-Hungarian nationalities within the state, whose aspirations had been ignored. In the Austrian half, the Czechs were left discontented, since they had historical justifications to claim special recognition equal to that advanced by the Hungarians, and the Austrian Germans had every reason to regret the loss of what they perceived to be their right to leadership in the larger empire. The Habsburgs turned over those nationalities inhabiting the Hungarian half to the not-so-tender mercies of the Magyars, who looked on their share of the empire as an essentially Magyar nation-state and acted accordingly.

Since the majority of the non-German and non-Hungarian nationalities fell within the borders of the Hungarian half of the state, the xenophobic Magyarization policies of the governing nationalists intensified the internal instability of the newly reshaped empire. The Croats, who had battled the Hungarians in 1848 and 1849 on behalf of the Habsburgs in vain hopes of winning imperial gratitude for their national cause, were particularly rankled. By cutting an expedient deal with one nationalist faction in their population and then committing themselves to preserving it at all costs, the Habsburgs created a plethora of increasingly strident nationalist movements that threatened to undermine from within the continued existence of the empire. Just like the 1848 revolutions, the compromise demonstrated that nationalism and liberalism were essentially incompatible in the best of circumstances and were dangerous for a multinational state.

THURINGIA SAXONY   PRUSSIA   RUSSIA

SILESIA

Sadova

Prague

K. of BOHEMIA

Cracow   *GALICIA*

BAVARIA

Danube R.

Munich

SLOVAKIA

BUKOVINA

Vienna

HUNGARY

AUSTRIA

Buda   Pest

SLOVENIA

Danube R.

Cluj

TRANSYLVANIA

Venice

Zagreb

Timișoara

K. of CROATIA

BANAT

ROMANIA

ITALY

DALMATIA

BOSNIA

OTTOMAN

Belgrade

SERBIA

Sarajevo

Danube R.

Split

Mostar

BULGARIA

*HERCE-
GOVINA*

**MONTE-
NEGRO**

EMPIRE

Dubrovnik

*SANDJAK OF
NOVI PAZAH*

	Austrian lands
	Hungarian lands
———	Border of Austria-Hungary
– – –	Border between Austria and Hungary

MILES

0   50   100   150   200

0   100   200   300

KILOMETERS

Map 35: The Balkans, 1878–1885

The outbreak of an anti-Ottoman revolt in Bosnia-Hercegovina in 1875 provided Serbia with an opportunity to shed Ottoman suzerainty entirely and commence territorial expansion. It also provoked terse European Great Power rivalries, known as the "Eastern Question." In 1876 Serbia, in conjunction with Serb-inhabited Montenegro, declared war on the Turks in aid of the rebels and in hopes of acquiring the two rebellious provinces. Russia supported the Serbs' action, prompting England to intensify its efforts to preserve intact what remained of the Ottoman Empire in Europe—to foil any increased Russian presence in the strategically crucial Balkans that might be gained by a Serbian victory. The Habsburgs sought to validate their Great Power status by dominating the Serbs and by expanding their empire into Bosnia-Hercegovina. They too could not afford to permit the Russians a strong foothold in the Balkans.

The international crisis seemed calmed when the Serbs were defeated. England called an international conference in Istanbul to force reforms and a new, Western-style liberal constitution on its Turkish allies. In May 1876, however, a rebellion occurred in the Ottoman Bulgarian province. It had been hastily prepared to take advantage of the Turks' preoccupation in the northwest. The uprising was a sad affair, badly planned and haphazardly undertaken. The ill-armed and disorganized rebels did little more than publicly rally, sing patriotic songs, and butcher their mostly pacific Muslim neighbors in the villages. The Turks organized swift retaliation and crushed the rebellion within a month. The brutality of the Turks' irregular forces resulted in at least 15,000 Bulgarian deaths and widespread devastation.

Word of these "atrocities" filtered outside the Bulgarian lands and eventually found its way into the Western news media. In England, it galvanized public opinion against Prime Minister Disraeli's pro Turkish policies and forced England to stand aside when Russia declared war on the Ottoman Empire in 1877 with the publicly proclaimed goal of winning for the Bulgarians a free nation-state of their own.

Although the Russo-Turkish war began somewhat inauspiciously, by the end of the year the Russians had crushed all Turkish military formations in the field. At the end of February 1878 Russian forces were in sight of Istanbul itself, and it seemed that Russia would finally realize its imperialist dream of acquiring the city and, with it, free access to the Mediterranean. But England reacted strongly against that turn of events, sending a fleet to the straits with orders to intervene should the Russians attempt to snatch Istanbul. Taking the hint, Russia halted and forced the Turks to sign the Treaty of San Stefano (March 1878), granting Serbia, Montenegro, and Romania (the Wallachian and Moldavian principalities had been united in 1856) complete independence, and creating a large Bulgarian puppet-state that dominated the central and eastern Balkans.

European reaction against the treaty was such that it became obvious the treaty needed modification to maintain the peace. The Great Powers met in Berlin in June-July 1878, and a new treaty was issued. All of the states that had received independence in San Stefano were confirmed; Russia acquired Bessarabia; Romania received a slice of Dobrudzha; and Austria-Hungary was permitted to occupy Bosnia-Hercegovina and the Sandjak of Novi Pazar, which separated Serbia from Montenegro and thus prevented their unification into a Great Serbian state. San Stefano Bulgaria was divided into four parts: an autonomous Bulgarian Principality, with an elected prince who governed under technical Turkish suzerainty; Eastern Rumelia, whose capital was at Plovdiv, with a Christian governor appointed by the sultan and approved by the Great Powers; Western Thrace and Macedonia were both returned to direct Ottoman control.

The Berlin settlement created dissatisfaction among the Balkan states. The dismemberment of Greater Bulgaria struck the Bulgarians like a hammer blow. National euphoria swiftly changed to disillusionment and then to stubborn resolution to win back that which had been lost. Bulgarian faith in the Russians was shaken, even though the new liberal-democratic Bulgarian government was shaped with Russian encouragement. Russians held most of the prominent positions in the principality's ministries, and the infant Bulgarian military was trained and officered by Russians. The Russian presence, however, proved incapable of stifling Bulgarian nationalism's dangerous momentum.

The Serbs, forced to accept Habsburg occupation of Bosnia-Hercegovina, also felt betrayed by the Russians, who appeared willing to give Bulgarians other regions important to Serb national aspirations. They made an accommodation with the Habsburgs, freeing their hands to deal with their Bulgarian rivals. The Greeks, whose national ambitions were ignored at Berlin, were resolved to make every effort to win what they considered their rightful borders in the north. The nationalist ambitions of all three of those Balkan peoples would collide violently in Macedonia. (See Map 36.)

In 1885 Bulgarian Prince Aleksandŭr I Battenburg (1879-86) unified Eastern Rumelia with the Bulgarian Principality despite the vehement protests of Russia, which withdrew all its advisors, ministers, and military officers from Bulgaria. In late 1885 Russia's actions prompted Serbia, which feared that nationalist momentum from the unification would carry the Bulgarians into Macedonia, to declare war on Bulgaria. The Serbs expected an easy victory and territorial acquisitions. Instead, the Bulgarians repulsed the Serb invasion forces at Slivnitsa and then invaded Serbia. Only Habsburg threats to intervene on behalf of the Serbs stopped the Bulgarian invasion, and a peace treaty was signed, in March 1886, on the eighth anniversary of San Stefano. The Bulgarian unification of 1885 was secured.

GERMANY

Danube R.

Vienna

AUSTRIA - HUNGARY

Budapest

Cluj

GALICIA

RUSSIA

Iaşi

BESSARABIA

Zagreb

Venice

Novi Sad

Belgrade

ROMANIA

Zadar

DALMATIA

BOSNIA -
HERCEGOVINA

Sarajevo

SERBIA

Bucharest

Danube R.

DOBRUDZHA

ITALY

Dubrovnik

SANDJAK

MONTENEGRO

Cetinje

Shkodër

Novi
Pazar

Niš

Vidin

Skopje

Sofia

Slivnitsa

BULGARIA

Tŭrnovo

Plovdiv

EASTERN
RUMELIA

Varna

Burgas

Naples

ALBANIA

Ohrid

MACEDONIA

OTTOMAN

Edirne

San Stefano

Istanbul

THRACE

Thessaloniki

EPIROS

Ioannina

Larissa

THESSALY

LIVADIA

Athens

EMPIRE

SICILY

Patras

GREECE

CYCLADES
ISLANDS

MILES

0   50   100   150   200

0   100   200   300

KILOMETERS

CRETE

Great Bulgaria

Borders in Treaty of San Stefano (3 March 1878)

Final border of Bulgaria, June 1878

Border of Bulgaria, 1885

Map 36: The Macedonian Question

Between the Congress of Berlin (1878) and the Balkan Wars (1912-13), nationalist political affairs in the Balkans were dominated by the "Macedonian Question." It was a conflict among Bulgaria, Greece, and Serbia—three states whose territorial aspirations had been disappointed or ignored by the European Great Powers meeting in Berlin—for possession of the Ottoman province of Macedonia, a region slightly larger than Vermont. At Berlin, the Russian-dictated San Stefano borders of newly recreated Bulgaria, which encompassed Macedonia, were overthrown, and the state was reduced to a fraction of the size Bulgarians considered acceptable. The Serbs had been forced to relinquish some territories won in the 1877-78 Russo-Turkish War and, more important, they were compelled to accept Austro-Hungarian occupation of Bosnia-Hercegovina, a region they adamantly claimed as Serb national territory. (See Map 38.) The Greeks, who had been restrained from participating in the recent war against the Turks, felt insulted by England's occupation of Cyprus, which they claimed as their own. All three turned toward Macedonia as compensation for their perceived losses at Berlin. (See Map 35.)

The Bulgarians advanced strong arguments supporting their claim. Macedonia had been an integral part of the first medieval Bulgarian state (681-1018), during which its regional capital of Ohrid had emerged as a leading Slavic cultural center and seat of the first independent Slavic Orthodox church, the Bulgarian archbishopric-patriarchate of Ohrid. During the latter decades of the state's existence under Tsar Samuil, Macedonia constituted the core of the Bulgarian state itself. (See Maps 10 and 12.) The language of the region's Slavs was so similar to Bulgarian that it was considered a dialect rather than a separate tongue. Some leading exponents of the 19th-century Bulgarian national revival, such as the Miladinov brothers, were from Macedonia. The Bulgarian's national movement evolved into the "Bulgarian Church Question" (1860-72), which succeeded in winning the Ottoman Turks' recognition of a Bulgarian *millet* through the institution of an autonomous Bulgarian church—the Exarchate—separate from the Orthodox *millet* controlled by the Greek patriarchate of Constantinople. Since the Turks decreed that any region where two-thirds of the population voted to join would fall under the authority of the new Bulgarian church, most Macedonian Slavs eventually voted to join the Exarchate, thus removing themselves from direct Greek ecclesiastical control.

Greek claims to Macedonia were grounded in allusions to ancient Macedon and its famous rulers, Philip II (359-36 B.C.) and Alexander the Great (336-23 B.C.) but were primarily focused on Byzantine possession of the region and later Greek control of the Ottoman Orthodox *millet*. Following independence from the Turks, nationalist Greeks concocted the Great Idea *(Megale idaia),* a political program calling for restoration of the Byzantine Empire as the natural Greek nation-state. The borders would be defined by territories in which the Greek language dominated within the Ottoman Orthodox *millet*. After the Turks disbanded the Bulgarian archbishopric-patriarchate of Ohrid (1767), Macedonia lay under direct Greek patriarchal control and a modicum of Hellenization occurred. To Greek nationalists, Macedonia was rightfully Greek. During the Bulgarians' campaigning to convince the Macedonians to join the Bulgarian Exarchate, the Greeks attempted to counter such activities with efforts of their own. The situation inside Macedonia quickly degenerated into Bulgarian-Greek violence and terrorism, with the native Macedonians becoming the primary victims.

Barred from expansion into Bosnia-Hercegovina, the Serbs were forced to look to Macedonia for possible future expansion. They too possessed historical claims on Macedonia. During the reign of Serbian tsar Stefan Dušan, his capital was the Macedonian city of Skopje and Macedonia formed the heart of his state's lands. Dušan's empire served as the territorial model for the modern state desired by the Serb nationalists. (See Map 18.) Failing to win lands from the Bulgarians by force in 1885-86, the Serbs felt compelled to enter the fray for possession of Macedonia. Serbian bands soon joined those of the Bulgarians and Greeks in the ethnic fighting that continued to plague and terrorize Macedonian natives.

In 1893, the Macedonian Slavs formed a revolutionary-nationalist organization of their own—the Internal Macedonian Revolutionary Organization (IMRO)—to fight against both the violent activities of the outsiders' nationalist bands and continued Ottoman control. Its members mostly were native Macedonians with some Bulgarians. IMRO's program was "Macedonia for the Macedonians," and, despite periodic dominance of its leadership by pro-Bulgarian elements within the organization, it signalled the emergence of a new, strictly Macedonian, nationalist movement.

Clashes among the nationalist parties in Macedonia became endemic, while attacks on Turkish authorities multiplied. In 1903 IMRO sparked an unsuccessful anti-Turkish uprising, which resulted in a futile intervention by the Western Great Powers to stop the bloodshed. Constant violence and terrorism uprooted thousands of Slav Macedonians, most of whom fled to southwestern Bulgaria, where they established a virtual state-within-a-state and became a militant force in Bulgarian politics. By 1912 all of the protagonists in the Macedonian struggle came to realize that the Turkish presence had to be eliminated before any further nationalist solution could be achieved. With Russia's urging, Bulgaria, Serbia, and Greece put aside their antagonisms long enough to form an anti-Ottoman alliance aimed at expelling the Turks from Europe and settling the Macedonian problem. The results were the Balkan Wars. (See Map 39.)

SERBIA

Vranje

BULGARIA

Prizren

Kyustendil

Plovdiv

Struma R.

Skopje

Kratovo

Gorna
Dzhumaya

Mesta R.

Vardar R.

Štip

Bansko

MACEDONIA

Nevrokop

Debar

Strumica

Nestos R.

Prilep

Ohrid

Resen

Strymon R.

Serres

Kavalla

Bitola

Axios R.

ALBANIA

Florina

Pella

Korçe

Kastoria

Thessaloniki

GREECE

Mount
Athos

Grevena

Konitza

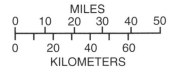

MILES

0   10   20   30   40   50

0   20   40   60

KILOMETERS

Map 37: The Balkans, 1908

A Turkish nationalist movement emerged in the Ottoman Empire in the 1860s when Westernized Turks, critical of traditional governing policies and opposed to superficial reforms espoused by the sultans, were forced to flee abroad. Establishing themselves in Paris, by 1902 they acquired the name of Young Turks and divided into two factions. One stood for centralizing the existing Ottoman state under Turkish predominance, the other called for decentralization and full ethnic autonomy for all of the empire's subjects.

While the Young Turks argued in Paris, officers of Turkish units in Macedonia acted. The movement, grounded in an empire-wide secret military officers' organization, the Society of Liberty, headquartered in Thessaloniki, espoused the ultranationalist centralization program of the extremist Young Turks. Well organized and supported by their troops, the Macedonian officers, led by Enver Pasha and including Mustafa Kemal, revolted in 1908 in response to the Western Great Powers' plan to intervene in Macedonia and halt the continued unrest there. Fearing a possible partition of the empire, the military leaders put a well-considered plan into operation. An ultimatum to implement the British-imposed reform constitution of 1876 that had never been put into force was telegraphed to Sultan Abdül Hamid II (1876-1909). (See Map 35.) The military units in Macedonia demonstrated their support, and the government was constrained to comply.

The jubilation of Westerners and non-Muslims in the empire turned to anger and fear as it became apparent that the Young Turks were intent on preserving the old empire as a Turkish nation-state. The revolutionaries, organized as the Committee for Union and Progress, subordinated the sultan to their will. They initiated a policy of centralization and Turkish hegemony formerly unknown in the Ottoman Empire and counter to the spirit of the constitution that they had ostensibly risen to instate. Virtually every non-Turkish subject population was forced to react against the new regime, spawning the nationalist awakenings of the Albanians, Arabs, and Armenians. The pseudo-Western ultranationalist program of the Young Turks ultimately led to atrocious massacres among the Turks' subject populations (such as the Armenians) and to the swift disintegration of the Ottoman Empire during and after World War I. (See Maps 41 and 42.)

The year 1908 also witnessed the outright annexation of Bosnia-Hercegovina by Austria-Hungary. It had implemented its Berlin right to occupy Bosnia-Hercegovina almost immediately after the treaty (1878). (See Map 35.) The occupiers initiated a civic works regime not seen in the Balkans since the French Illyrian episode. (See Map 30.) By all Western standards, Habsburg occupation should have been a boon for the inhabitants. Unfortunately, such was not the case. The Habsburgs mistakenly failed to dismantle the old Muslim-controlled landholding regime that had evolved there

over the centuries. A few thousand Muslim *beys* continued on as large estate owners and wielded immense local power over tens of thousands of Christian peasants. The estates were run in obsolete fashion, with inefficient methods of land use and outmoded equipment and techniques. Thus agriculture, the region's primary economy, remained backward and the population poor. Despite all the other benefits that Austrian occupation brought to Bosnia-Hercegovina, the majority of the population—Christians of both stripes—remained downtrodden and grew increasingly discontented. Their unrest often found release in revolutionary activity against the powerful Muslim oligarchy and continued Austrian occupation.

The 1908 Young Turk Revolution threw Europe into turmoil, and the Great Powers hastily convened foreign policy meetings among themselves in attempts to reshuffle the balance of power in the Balkans should the Ottoman state fall apart completely. In September, such a meeting between the foreign ministers of Austria-Hungary and Russia took place in Buchlau, Austria, at which the Austro-Hungarian foreign minister Count Alois Aehrenthal finessed his Russian counterpart, Alexander Izvolsky, into accepting the Habsburgs' outright annexation of Bosnia-Hercegovina in exchange for empty words regarding future Austrian support for Russia's claim to free access to the Mediterranean—essentially nothing. In October 1908 Austria-Hungary announced its intention to annex Bosnia-Hercegovina permanently.

The announcement was met by strongly voiced concerns by the Great Powers and by frenzy and rage among the Serbs. The Russians protested that they had been duped. Germany supported its Habsburg ally, while the French and English stood by Russia. None acted, however, since all feared that a general European war would result (given the Central Powers–Entente alliance system then in place). In 1909 the Turks accepted the annexation in return for compensation from Austria-Hungary, and the crisis ended.

Taking advantage of the Great Powers' preoccupation with the annexation crisis, Bulgarian Prince Ferdinand I (1887-1918) declared Bulgaria completely independent of Ottoman suzerainty, which freed him to pursue ambitions in Macedonia. A bloody military coup in Belgrade in 1903 had overthrown the pro-Habsburg Serbian Obrenović dynasty and installed the Russian-looking Karadjordjevićes, but Russia's problems with Japan forced the new Serbian ruler Petr I (1903-21) to attempt an accommodation with Bulgaria to resist mounting Habsburg hostility. The nationalist and anti-Habsburg policies of Nikola Pašić, Petr's prime minister, provoked a tariff conflict with Austria-Hungary known as the "Pig War" (since Serbia's primary exports were pork products, and Austria-Hungary was the chief market). The rapprochement between Serbia and Bulgaria was short-lived, as they floundered over the issue of Macedonia, and seemed doomed once Ferdinand made his move for independence.

GERMANY

AUSTRIA - HUNGARY

Danube R.

Vienna

Budapest

GALICIA

RUSSIA

Iaşi

BESSARABIA

Zagreb

Novi Sad

Belgrade

Cluj

ROMANIA

Bucharest

Danube R.

DOBRUDZHA

Venice

Zadar

DALMATIA

BOSNIA - HERCEGOVINA

Sarajevo

SERBIA

SANDJAK

Novi Pazar

Niš

Tŭrnovo

Varna

ITALY

Dubrovnik

MONTENEGRO

Cetinje

Prizren

Skopje

Sofia

Plovdiv

BULGARIA

Burgas

Naples

Shkodër

ALBANIA

Uhrid

MACEDONIA

Thessaloniki

OTTOMAN

Edirne

THRACE

Istanbul

EMPIRE

EPIROS

Ioannina

Larissa

SICILY

THESSALY

Athens

Patras

GREECE

CYCLADES ISLANDS

MILES

0   50   100   150   200

0   100   200   300

KILOMETERS

CRETE

Bosnia-Hercegovina annexed by Austria-Hungary

Sandjak of Novi Pazar occupied by Austria-Hungary

Map 38: Bosnia-Hercegovina, 1908–1914

In 1878, when Bosnia-Hercegovina was occupied, the population was divided among three component elements. The largest single element was the East European Orthodox (some 43 percent), followed by the Muslim (39 percent), and finally the Western European Catholic (18 percent). Both of the Christian components had developed ethnonational self-identities through influences that had infiltrated into Bosnia from its neighbors—the Orthodox espoused a Serb identity and the Catholics a Croat. Given the traditional theocratic culture of Islam, the Muslims, though ethnically Slavic and speaking the same language as the Christians, held no ethnonational affiliation. They maintained an Islamic cultural self-identity alone.

In both Croatia and Bosnia-Hercegovina, Croat and Serb nationalists' reaction to the 1908 annexation was strong. The more moderate, traditional Croat Yugoslav nationalists saw it as opening a bright future, in which their dreams for creating an autonomous Croat-led Yugoslav state within the Habsburg Empire would be fulfilled. The heir to the Habsburg throne, Archduke Francis Ferdinand, had made it known that he was considering favorably the idea of restructuring Austria-Hungary once he attained power by extending to Czechs and, possibly, Croats political autonomy similar to that enjoyed by the Hungarians. Radical Croat nationalists considered the annexation the first step on the road toward a Greater Croatia, but, for the same reasons, a recently founded Serbo-Croat coalition considered it a catastrophe. To them, Francis Ferdinand's trialistic sympathies for a tripartite state system marked him as the personification of the Austro-Hungarian threat to their future aspirations for complete separation from the Habsburgs. Nurtured by Serbia, the Serbo-Croats bewailed the annexation as an insult to Yugoslavism (by which they meant, whether the Croat partners realized it or not, Serb nationalism). Repression by Hungarian administrators in Croatia succeeded only in discrediting Austro-Hungarian administration in the eyes of most Croats and Serbs and intensifying the discontent of the Belgrade-looking faction.

By 1909 in Bosnia-Hercegovina, a new element that was dangerous to Austro-Hungarian rule in the region began to make itself felt among the Serb population. In 1902 a cultural society called *Posveta* (Enlightenment), funded in part by money from Serbia, was established for the express purpose of educating peasant and lower-class Serb children. Within a decade of its founding, the society spawned a new type of Bosnian Serb intellectual—poor, often jobless, with no vested interest in the Austrian-imposed establishment (the Habsburg administration tended to hire Croats over Serbs), and with a chip-on-the-shoulder attitude toward the inequitable existing social system.

That younger generation of Bosnian Serbs formed the cadres of a movement known as "Young Bosnia," an amorphous but widespread association that sought independence from the Habsburgs and social reforms within a Bosnian Serb nation-state. There was little agreement among its members as to how their objectives were to be attained, but a general affinity among them for Russian revolutionary literature led them to follow mostly Russian models. Turning their backs on evolutionary political reform tactics, the Young Bosnians embraced terrorism, which they elevated into a veritable cult. Terrorist acts and the "martyrs" that such actions invariably created inflamed their blood and inspired their efforts. By 1912 members of Young Bosnia were in direct contact with a secret Serb ultranationalist revolutionary organization, commonly known as the "Black Hand" but correctly named "Union or Death" *(Ujedinjenje ili Smrt)*. The Black Hand was controlled by Serbian military officers holding high positions in the Belgrade government—the very officers who had conspired to kill King Aleksandr Obrenović in 1903—and led by Colonel Dragutin Dimitrijević, also known as "Apis." While King Petr disliked the Black Hand leadership personally (they were, after all, regicides) and they, in turn, operated beyond his control and often at variance with his policies, he tolerated the organization's existence because of its anti-Habsburg and Greater Serbia stance. Through its Young Bosnia contacts, the Black Hand engineered and armed terrorist activities inside Bosnia.

In the five years preceding the outbreak of World War I in 1914, the Habsburg authorities in Bosnia came down hard on Young Bosnian agitation, making hundreds of arrests for treason and espionage and mostly winning convictions. In turn, the youthful Bosnian Serb revolutionaries intensified their violent actions. When in the early summer of 1914 it was announced that Archduke Francis Ferdinand would make a tour in Bosnia and would visit Sarajevo on 28 June, Vidovdan (St. Vitus' Day)—the anniversary of the Battle of Kosovo Polje (1389), at which medieval Serbia went down in bloody defeat to the Turks, and which Serb nationalists considered sacred in a morbid sort of way ("Remember our defeat so that we will never let it happen again")—neither the Young Bosnians nor the Black Hand could let pass the opportunity to murder that personification of the Habsburg threat to Greater Serb national aspirations. In a comedy of errors that would have been humorous if it had not been so fatefully tragic, the Young Bosnian tool of the Black Hand, Gavril Princip, managed to shoot the Habsburg heir to the throne (and his wife). Princip's shots on a Sarajevo streetcorner and the two bodies of his victims proved to be but the first of a thunderous barrage and millions of corpses that spanned four years, as the Central Powers–Entente alliance system refused to permit Austria-Hungary to punish Serbia for the Sarajevo crime by means of a limited, localized military drubbing. Within a month of the outbreak of the Habsburgs' war of retribution, begun in July 1914, the struggle mushroomed into all-out world war. (See Map 41.)

AUSTRIA

Ljubljana

SLOVENIA

Rijeka

CROATIA

Zadar

DALMATIA

Šibenik

Split

Varaždin

Zagreb

HUNGARY

Drava R.

SLAVONIA

Sava R.

Osijek

Vukovar

Danube R.

VOJVODINA

Novi Sad

Bihać

BOSNIA

Banja
Luka

Jajce

Travnik

Zenica

Bosna R.

Tuzla

Sarajevo

Srebrenica

Drina R.

SERBIA

Belgrade

Višegrad

Neretva R.

Mostar

HERCE-
GOVINA

Drina R.

SANDJAK

Novi Pazar

MONTENEGRO

KOSOVO

Dubrovnik

Kotor

Cetinje

ALBANIA

MILES

0 10 20 30 40

0 20 40 60

KILOMETERS

Bosnia-Hercegovina

Border between Austria and Hungary

Occupied by Austria-Hungary until 1912

Sandjak and Kosovo were part of the Ottoman Empire

Map 39: The Balkan Wars, 1912–1913

The repressive Turkish nationalist policies of the Young Turks played into the hands of the Balkan nationalists, eventually permitting them to overcome briefly their mutual animosities and to form an anti-Turkish military alliance in 1912. (See Map 37.) They were encouraged by Russia. Knowing that Russia supported a Serb-Bulgarian alliance, and taking advantage of the Turks' involvement in a war with Italy over Tripoli (1911), Serbia and Bulgaria hammered out a military treaty of mutual assistance in early 1912. A secret annex dealt with the future fate of Balkan regions still under Turkish control. The Sandjak of Novi Pazar, Kosovo, and a large strip of northern Macedonia were to go to Serbia. Western Thrace was ceded to Bulgaria. The bulk of Macedonia was to be formed into an autonomous province. Should the autonomous province prove unworkable, the secret provision provided for its further division, with Bulgaria and Serbia each receiving additional territories and the remaining areas subject to Russian arbitration as to their final allotment. Soon thereafter, a Greek–Bulgarian, anti-Turk military alliance was signed, in which no territorial issues were defined since both states desired the important Macedonian port of Thessaloniki, and Montenegro signed alliances with both Serbia and Bulgaria. Thus the Balkan League attained its final composition.

Meanwhile, the Ottoman state was harried by Italian attacks and in internal disarray. In May 1912 the Albanians had risen against the Young Turks, and Ottoman military morale and strength were collapsing. By October 1912, conditions were ripe for the Balkan allies to begin their war on the Turks.

Ignoring Russian pleas to wait, Montenegro declared war on 7 October, followed ten days later by the other Balkan allies. There was little doubt that the war was fought primarily to decide the fate of Macedonia. But geography forced the Bulgarians, the easternmost of the allies, to focus their efforts in the wrong direction, against the main Turkish forces in Thrace, while their three allies faced mostly demoralized and understrength enemy units in the west, in and around Macedonia. Serb forces easily overran and occupied close to two-thirds of Macedonia and then invaded Albania. Greek troops pushed into Epiros and southern Macedonia, occupying Thessaloniki over loud Bulgarian protests. The Bulgarians found themselves in a bloody contest for Edirne and received a gruesome foretaste of trench warfare in their assaults on successive Turkish fortified positions at Lüleburgaz and Çatalca. By the time Edirne fell to the Bulgarians in March 1913, only Istanbul itself and Shkodër, in Albania, remained of Turkey-in-Europe. In April an armistice was signed. In May the Treaty of London ended hostilities.

Dissention soon arose among the victorious Balkan allies over the disposition of conquered territories. The European Great Powers decided to create an autonomous Alba-nia, which included areas originally ceded to Serbia in alliance treaties. In compensation, the Serbs demanded a larger share of Macedonia, to which the Bulgarians adamantly objected. Both the Bulgarians and the Greeks were at loggerheads over possession of Thessaloniki. Smelling nationalist blood, the Romanians, who had remained neutral during the war, placed a bid for southern Dobrudzha, which had been in Bulgarian hands since 1878. Russia attempted to smooth the frictions among the allies but failed. In June 1913 Serbia and Greece concluded an anti-Bulgarian alliance to defend their zones of occupation in Macedonia against possible Bulgarian encroachment, and then won Montenegrin support. A Russian mediation initiative was ignored.

All three of the contentious allies transferred troops to the lines established in and around Macedonia. Border clashes between Bulgarians and their now belligerent allies multiplied. Nationalist emotion in Bulgaria built to a feverish pitch. The army grew restless and demanded action or demobilization. Bulgarian public opinion, whipped up by the agitation of various Macedonian immigrant groups who threatened to assassinate Ferdinand and important members of his government if they did not act, clamored for war against both Greeks and Serbs. The military high command, which had hurriedly redeployed the bulk of the army from the eastern front facing the Turks to the western front facing Macedonia, assured Ferdinand that all was ready for decisive action. In late June the Bulgarians attacked Serb and Greek positions in Macedonia. It was a naive and foolish move.

Serbia and Greece immediately declared war on Bulgaria. Montenegro followed, and in July both Romania and the Ottomans did likewise. The Bulgarians found themselves in an untenable military position and could offer only meager resistance to the concerted attacks of their enemies. They were easily defeated by the Serbs and Greeks in Macedonia, while the Turks regained most of Thrace up to and including Edirne, and the Romanians captured southern Dobrudzha. In a little over a month the Second Balkan War was over. By maneuvering the Bulgarians into playing the role of aggressors, the anti-Bulgarian allies had succeeded in ensuring that Bulgaria had forfeited any sympathetic support from the international diplomatic community. Bulgaria was stripped of most gains won in the first war, including Western Thrace and its port of Kavalla, Edirne and most of Eastern Thrace, and most new acquisitions in Macedonia, except for a slice in its northeast. Romania retained much of Dobrudzha, while Greece and Serbia divided the rest of Macedonia between themselves—the Greeks retaining Thessaloniki and the southern portions of the region; the Serbs acquiring the lion's share of the central and northern portions, including Bitola. The borders established for Bulgaria and Greece in 1913 proved relatively permanent into the present day.

BORDERS BEFORE THE FIRST BALKAN WAR, 1912

AUSTRIA - HUNGARY

Venice
Zagreb
Zadar
DALMATIA
BOSNIA - HERCEGOVINA
Sarajevo
Belgrade
SERBIA
ROMANIA
Bucharest
Silistra
Constanța
DOBRUDZHA
Danube R.
MONTENEGRO
Dubrovnik
Cetinje
SANDJAK
KOSOVO
Niš
Varna
BULGARIA
Sofia
Burgas
ITALY
Shkodër
ALBANIA
Skopje
MACEDONIA
Plovdiv
Naples
Durrës
Strumica
Edirne
Lüleburgaz
Bitola
Kavalla
THRACE
Çatalca
Vlorë
Thessaloniki
Enez
Istanbul
EPIROS
Florina
Ioannina
Larissa
OTTOMAN EMPIRE
GREECE
Athens
SICILY
Patras
CYCLADES ISLANDS
DODECANESE IS.

MILES
0   50   100   150   200
0   100   200   300
KILOMETERS

BORDERS AFTER THE FIRST BALKAN WAR, 1912

BOSNIA - HERCEGOVINA
SANDJAK
Belgrade
ROMANIA
Bucharest
Danube R.
DOBRUDZHA
MONTENEGRO
Niš
SERBIA
BULGARIA
ALBANIA
KOSOVO
Skopje
Sofia
MACEDONIA
Edirne
Vlorë
Strumica
THRACE
Istanbul
GREECE
OTTOMAN EMPIRE
Athens

MILES
0   50   100
0   150
KILOMETERS

BORDERS AFTER THE SECOND BALKAN WAR, 1913

BOSNIA - HERCEGOVINA
Belgrade
ROMANIA
Bucharest
SERBIA
SANDJAK
Danube R.
DOBRUDZHA
MONTENEGRO
Niš
ALBANIA
KOSOVO
Skopje
BULGARIA
Sofia
MACEDONIA
Strumica
Edirne
Vlorë
THRACE
Istanbul
EPIROS
GREECE
OTTOMAN EMPIRE
Patras
Athens

MILES
0   50   100
0   150
KILOMETERS

Map 40: Eastern Europe, 1914

German chancellor Bismarck was aware that the Congress of Berlin had intensified Habsburg-Russian animosities to the point of potential conflict. His first response, in 1879, was to forge an alliance between Germany and Austria-Hungary. He then attempted to shape a more general alliance of the three emperors of Germany, Russia, and Austria-Hungary (1881), who found common ground in their shared possession of Polish lands. Yet tensions continued. Both Germany and Russia treated their Polish populations badly, while the Habsburgs, forced to make concessions to liberalism and nationalism by the Hungarians, took the opposite approach and thus increased their frictions with Russia, since Habsburg Galicia became a "home away from home" for Polish and Ukrainian nationalists intent on throwing off Russian domination. The growing liberalism of Austria-Hungary led to the downfall of the Three Emperors' League. Bismarck had realized its fragility from the start—in 1882 he concluded the Central Alliance linking Germany, Austria-Hungary, and Italy.

When Bismarck was expelled from office in 1890 by Kaiser William II (1888-1918), German-Russian foreign policy alliances collapsed, and Russia, needing outside investment for its tardy industrialization, moved closer to France, which needed political support against Germany. The 1894 secret alliance between the two powers finalized the web of international treaties that divided the European states into two opposed camps: the Austro-German (the Central Powers) and the Franco-Russian (the Entente). Italy and England, on the peripheries of the alliances, were eventually drawn into the system on opposite sides—Italy with Germany; England with France and Russia.

The alliance system might have worked had it been restricted to the Great Powers. But lesser states were drawn into the system. For example, Russia, having lost most of its direct influence in the Balkans at Berlin, turned to Serbia as a Balkan ally. This relationship strengthened when, instead of weakening Serb nationalistic ambitions, the Habsburgs' occupation of Bosnia-Hercegovina intensified them and created an intractable Serb enemy. After a coup in 1903, in which the pro-Habsburg Serb king Aleksandr Obrenović (1889-1903) was butchered by his military officers and the pro-Russian Petr I Karadjordjević (1903-21) installed in his place, Serbia moved firmly into the Russian camp.

Habsburg emperor Francis Joseph was not a national ruler of a national empire. Instead, Austria-Hungary was merely the representation of a dynasty that crossed all territorial-national boundaries within its borders. Only loyalty to the House of Habsburg served to cement the disparate parts of the empire together. The various ethnonational groups in the state looked to Francis Joseph for satisfaction of their rising national aspirations, but his hands were tied by commitments to his Hungarian partners and intricate constitutional matters that governed relationships between himself and the various peoples under his rule.

In the Austrian half of the empire, Habsburg problems were mostly caused by the Czechs, who claimed historical constitutional rights rooted in the medieval Bohemian state (the "Crown of St. Vaclav"). But the legacy of the Battle of White Mountain (1620) killed Czech chances of becoming a "Second Hungary" by reducing Czech constitutional rights. (See Map 27.) Unlike the Hungarians, who had won a large measure of constitutional independence through past armed resistance to the Habsburgs, the Czechs had remained mostly passive and emerged in the 20th century in a subordinate position. Faced with assimilation into a German-dominated Austrian half of the empire, the Czechs strove for policies expanding the limited rights of the Bohemian crown into a constitutionally autonomous and relatively separate Bohemia similar to Hungary after the *Ausgleich*. (See Map 34.) The emperor could not satisfy Czech demands without endangering the position of his Hungarian partners, whose national minorities most likely would demand similar compensation.

In the Hungarian half of the empire, national matters were far more serious. The Magyars were utterly committed to preserving the status quo of the *Ausgleich* that gave them exclusive national rights and control over their area. They pursued an anti-minority policy that often crossed over into outright persecution. In Transylvania the predominantly peasant Romanian population was weak and divided. They looked not to Romania but to the emperor in Vienna to find support for their national aspirations. Never having possessed a historical state of their own to use as a precedent, the Romanians' hopes were more futile than those of the Czechs.

Even more so were those of the Slovaks, the least influential minority in the Hungarian half of the empire. For nearly a thousand years the Slovaks had been docile peasant subjects of the Magyars. Those Slovaks who had risen to prominence had done so by acculturating themselves to the Magyars. The Slovaks possessed vague cultural and historical ties to the Czechs, but these complicated any real attempts to forge closer ties. The Slovaks were the most culturally persecuted of all the minorities in the Hungarian sphere.

The Croats posed the chief national minority problem for the Hungarians. They had been forced to make a constitutional deal with the Hungarians after the *Ausgleich*. The following year a compromise was reached: The Croats accepted continued Magyar rule in return for limited representation in the Hungarian diet and Croat local administration in Croatia. Hungarian nationalists continued to demand direct Magyar control of Croatia, while Croat nationalists pressed for complete independence, holding up, as did the Czechs, the 1867 Compromise as a model for creating a tripartite empire. The 1908 annexation of Bosnia-Hercegovina served only to heighten their nationalistic demands.

SMOLENSK

LITHUANIA
Vilnius · Smolensk ·

Danzig · Königsberg ·
EAST PRUSSIA

Berlin ·   Minsk ·

Posen ·

GERMANY   RUSSIA

Warsaw ·
Breslau ·   POLAND

Prague ·   Kiev ·
BOHEMIA   Cracow ·   L'viv ·   UKRAINE

Danube R.
Brno ·
Munich ·   SLOVAKIA   GALICIA

Vienna ·   Bratislava ·

AUSTRIA-
Budapest ·   Iaşi ·   Odessa ·
HUNGARY   BESSARABIA

HUNGARY   Cluj ·

Zagreb ·   TRANSYLVANIA
Trieste ·
Venice ·   CROATIA
DALMATIA   Danube R.

BOSNIA-
HERCEGOVINA   Sarajevo ·   ROMANIA

Belgrade ·   Bucharest ·

Dubrovnik ·   SERBIA   Danube R.
Rome ·   Kotor ·   MONTENEGRO   Niš ·
Cetinje ·   Sofia ·
Shkodër ·   BULGARIA
Naples ·   ALBANIA   Skopje ·

ITALY   Edirne ·
Elbasan ·   Thessaloniki · Kavalla ·   Istanbul ·

GREECE   OTTOMAN
EMPIRE

SICILY   Izmir ·

Athens ·

MILES
0   50 100 150 200

0   100 200 300
KILOMETERS
DODECANESE IS.
To
Italy
CRETE

- - - - Boundaries of Kingdom of Hungary

Map 41: Eastern Europe during World War I

World War I came as an undesired accident. All of the Great Powers tried to head it off, yet war erupted. Austria-Hungary wanted to squelch the threat of Serb minority nationalism by thrashing Serbia decisively in a localized war. Once Habsburg troops invaded Serbia in July 1914, however, that proved impossible because the weblike system of alliances precluded all efforts to prohibit the proliferation of military action. Russia's alliance with Serbia demanded military mobilization against Austria-Hungary. Once that occurred, Germany was obliged to mobilize in support of Austria-Hungary, which, in turn, forced France to follow suit according to its treaty with Russia. Unfortunately, military mobilization meant that combat could not be averted. Within a little over a month of the tragedy in Sarajevo, the European world found itself embroiled in total war.

The conflict was never confined to the Balkans, and Austria-Hungary never achieved swift victory over Serbia. The quick knockout-blow victories sought by all the other participants once the fighting became general in Europe never materialized. The war settled into a grueling and costly conflict, in which ultimate victory went to the side that could best survive the astronomical costs in human lives, material resources, and socioeconomic pressures charged by the fighting.

Austria-Hungary's initial conflict with Serbia dragged on into 1915, with a number of humiliating reverses. Only the intervention of Bulgaria on the Serbs' right flank and rear, after the Central Powers bribed it out of initial neutrality with a grant of all Macedonia, drove the Serbs over the Albanian Alps to the Adriatic, from which the British evacuated them first to Corfu and then to Thessaloniki. Bulgarian and Habsburg troops then occupied all of Macedonia and Serbia, respectively. Meanwhile in the north, operations in 1914 opened with a Russian invasion of Germany that swiftly ended in crushing defeat at Tannenberg. Thereafter, costly attacks and counterattacks between Russian and German-Habsburg forces in Poland and Galicia proved indecisive throughout the rest of the year.

The Ottoman Empire joined the Central Powers in July 1914 and proceeded to attack Russia in the Caucasus and England in the Middle East. England retaliated in 1915 by landing an invasion force on the Gallipoli Peninsula in an effort to capture Istanbul and knock the Turks out of the war and to open a maritime supply route to its Russian ally. Stubborn Turkish defensive actions foiled the attempt, and the British were forced to evacuate their troops in early 1916. Minor advances in Poland and Galicia were gained by the Central Powers in 1915 at enormous costs to both sides. The next year, Russia succeeded in pushing back Habsburg forces some fifty miles at exorbitant cost, resulting in widespread demoralization and discontent among Russian troops.

Following Italy's defection from the Central to the En-tente Alliance in 1915, a military front against Austria-Hungary opened in its alpine northeast (1916). There followed a series of battles characterized by military ineptitude common to both sides. On the eastern front against the Russians, Austria-Hungary fared little better. The empire's troops needed German bolstering to preserve their tactical integrity. In the Balkans, the Entente bullied Greece into permitting their forces to land at Thessaloniki and establish a front against the Bulgarians, Habsburgs, and Germans, who were advancing from the north. The Serb army was shipped there from Corfu and joined French, Italian, and a few Russians in the lines formed in Greek Macedonia. Attempting to take advantage of Habsburg preoccupation on the far-flung war fronts, Romania declared war on Austria-Hungary and invaded Transylvania (August 1916), only to be defeated and occupied in early 1917.

By 1917 Habsburg emperor Charles I (1916-22), who succeeded Francis Joseph, was aware of the disaster for his empire that loomed on the horizon should the Central Powers be defeated. His efforts to pull the empire out of the war before it was too late were stymied by his German allies and by his enemies' growing intransigence to accepting anything less than total victory over the Central Powers as a whole. His foresight was accurate. In April the United States entered the war on the Entente's side, and President Woodrow Wilson issued his fourteen-point declaration of American war aims, which called for the creation of a restored Polish state and for the right of nation-state national self-determination for all ethnonational groups within the borders of the Central Powers. Wilson's Fourteen Points served to undermine the morale of the polyglot peoples fighting in the Austro-Hungarian ranks.

By early 1917, Russia was in worse shape than Austria-Hungary. Huge losses in men and material, a dysfunctional rail transportation system that starved both army and cities of food and other supplies, and no end to hostilities in sight resulted in two revolutions (March and November) that overthrew tsardom and brought Vladimir Lenin and his Bolsheviks to power. Lenin, desperate to end the war so as to consolidate his political victory, agreed to a humiliating peace with Germany at Brest-Litovsk (March 1918), ending the war on the eastern front. In the Balkans, an all-out Entente offensive led to the rapid collapse of the German and Bulgarian forces facing them in September 1918. In a matter of fifteen days the front disintegrated, and Bulgaria signed an armistice ending the fighting. Soon thereafter Austria-Hungary collapsed as the various minority ethnonational groups in the empire (such as Czechs, Slovaks, Croats, Slovenes, and Romanians) declared their independent national existence along the lines of Wilson's Fourteen Points (October-December 1918). The Hungarians and Austrian Germans were left no alternative but to do likewise.

LITHUANIA	
Vilnius	
Tannenberg	Minsk
Berlin	RUSSIA
GERMANY	Brest-Litovsk
Warsaw	
POLAND	
Prague	Kiev
GALICIA	UKRAINE
Danube R.	
Vienna	BESSARABIA
Budapest	
AUSTRIA-HUNGARY	TRANSYLVANIA
Trieste	
Venice	ROMANIA
Rijeka	Belgrade
Sarajevo	Bucharest
BOSNIA-HERCEGOVINA	SERBIA
Danube R.	
ITALY	MONTENEGRO
Rome	Sofia
ALBANIA	BULGARIA
Skopje	
Bitola	Istanbul
Thessaloniki	
CORFU	Gallipoli
GREECE	OTTOMAN EMPIRE
SICILY	
Athens	Izmir
CRETE	

MILES
0 50 100 150 200

0   100   200   300
KILOMETERS

- – - – Front at end of 1914
- • • • • Front at end of 1915
- — — Front at end of 1916
- ——— Brest-Litovsk Treaty Line, March 1918
- ——— Armistice Line of September 1918

Part V

Modern and Contemporary Periods

(1918–1991)

Map 42: Eastern Europe, 1923

The political map of modern Eastern Europe originally was drawn by the victorious Entente Powers (specifically, Britain, France, and the United States) during negotiations at Versailles (1919-21) that ended World War I. No matter that their cartography was based on their own political and economic self-interests, once the borders were drawn they became political realities because their Western creators had the muscle to enforce their decisions. The Versailles configuration of the world was institutionalized by the founding of first the League of Nations and then, following World War II, the United Nations. These organizations were intended to legitimize the Western-imposed political settlement of states that had been shaped in the image of the West's nation-state principles and made state borders inviolable through a set of international laws.

In reality, Versailles Eastern Europe never reflected the Western ideal that each nation-state would truly represent the interests of its total population. The victorious Western architects of post–World War I Europe cynically used the doctrine of national self-determination in their mapmaking. The catastrophic results of Versailles regarding the victors' mistaken efforts to punish the Germans for the Great War are well known. By redrawing the borders of Germany in such a way as to cripple German industrial potential and to limit Germany's demographic base, the Versailles settlement violated the West's own avowed nation-state ideals and furnished the Germans with valid national grievances within the very context of Western European political culture.

But the obvious German example was merely a single case of the violation of principles of national self-determination at Versailles. Western recognition of state borders in Eastern Europe could not be separated from the victors' short-sighted policy of rewarding those regarded as allies and punishing those viewed as wartime enemies. Nowhere in geographical Eastern Europe did the national territorial claims of any one people go uncontested. The Versailles mapmakers resorted to public polls (plebiscites) to adjudicate border disputes among "friends" in some kind of seemingly equitable fashion, but similar conflicts between allied and "enemy" peoples were almost always judged in favor of the former, often with their most grandiose territorial pretensions satisfied. The borders of the winners—Poland, Czechoslovakia, Yugoslavia, Romania, and Greece—were determined at the expense of equally valid, but disregarded, claims of nation-state losers—Germany, Hungary, and Bulgaria.

Even if the mapmakers had been objective in their determinations, ethnonational development in Eastern Europe made the drawing of truly national borders difficult, if not impossible. Within the borders of the multinational Habsburg, Russian, and Ottoman empires, there had been few internal boundaries, which resulted in widespread territorial mixing of different societies over long periods. But the mapmakers were consciously unobjective in their work, and they openly admitted as much when, primarily as a public relations ploy, they established a League of Nations commission in Geneva to arbitrate the just but disregarded claims of the many national minorities created by their decisions.

Nationalism and the nation-state represented the West's rejection of artificially and arbitrarily constructed countries that were a legacy of the Middle Ages. They were declared "unnatural," since they combined different nationalities within common borders merely for the benefit of an elite governing class. In his Fourteen Points, Woodrow Wilson had proclaimed this as a primary war aim of the United States. Wilson's declaration served to inflame the national aspirations of numerous nationalities within the borders of the enemy Central Powers (thus helping undermine their military potential) and to force his allies to accept national self-determination of peoples as the avowed basis for the later Versailles peace. Wilson, like most Americans at the time, was naive when it came to high-level international politics, and was easily circumvented by his more experienced and pragmatic colleagues at the peace conference, who did not share his idealistic scruples regarding the nature of the nation-state when it came to furthering their own national agendas. At Versailles, the old anational political structures were dismantled, but, unfortunately, the new states were constructed more to ensure both the prolonged punishment of the defeated Central Powers and the political and economic interests of the Western Great Powers in Eastern Europe than to uphold Wilsonian ideals of national self-determination. Most of the new states artificially encompassed disparate nationalities: Czechoslovakia (Czechs, Slovaks, Germans, Hungarians, and Ruthenians); Yugoslavia (Serbs, Croats, Bosniaks [Bosnian Muslims], Bulgaro-Macedonians, Albanians, Hungarians, and a smattering of others); Romania (Romanians, Hungarians, Germans, Bulgarians, and Turks); and Poland (Poles, Germans, Lithuanians, Belorussians, and Ukrainians). (See Maps 43-48.)

The victors at Versailles were cognizant of the awkward national makeup of those new states. The original names of two were hyphenated, either literally or figuratively: Czecho-Slovakia and the Kingdom of the Serbs, Croats, and Slovenes (later, in 1929, Yugoslavia). After Versailles, and virtually until the post-1989 disintegration of those two states occurred, the West stubbornly refused to admit publicly the contradiction their recognition posed to the essence of Western European nation-state political culture by glibly speaking of "Czechoslovaks" and "Yugoslavs" as if such creatures existed as authentic nationalities. By placing their own political interests into the forefront of their mapmaking, the Western Great Powers at Versailles sanctioned new states as arbitrary in national makeup as the states they dismantled in the name of the nation-state principle.

Map 43: Independent Poland, 1920–1922

By the end of 1918, all three of the partitioning powers had fallen. Germany was exhausted by the war. Austria-Hungary disintegrated. Russia dissolved in a vicious civil war between Lenin's Bolshevik government and an assortment of anti-Communist and pro-tsarist forces. As these situations evolved, Józef Piłsudski, leader of the pro-German and anti-Russian Polish Socialist Party, managed to hammer together a political arrangement with his political rival, Roman Dmowski, head of the pro-Russian National Democratic Party, that aimed at taking maximum nationalist advantage of the prevailing conditions to found a new Greater Polish nation-state.

In November 1918 a German-founded and -supported Polish Regency Council declared war on Ukraine, which, in seeking to win independence from Soviet Russia, had invaded Galicia. Within days of that declaration, the regency proclaimed an independent Polish Republic and Piłsudski was given full military power. The regency then promptly resigned. The Poles cleared Galicia of Ukrainians and made additional westward advances, capturing Poznań and Great Poland from the Germans by year's end. Piłsudski founded a coalition government for the new Poland early in 1919, with himself installed as provisional president.

New Poland immediately set about conducting a policy of military expansion into Lithuania, Belarus, Ukraine, and Silesia to win control of all prepartition Polish territory before the Versailles conference could draw the official map of postwar Europe. The Poles planned to present it with a fait accompli. Yet despite the Poles' efforts at territorial aggrandizement, the Versailles mapmakers issued a dictate in late 1919 drawing Poland's new eastern border—the so-called Curzon Line—farther west than the Poles desired. Piłsudski, elevated to marshal and chief of state, refused to accept the Versailles settlement, since the Curzon frontier did not include Vilnius or most of western Ukraine and Belarus.

Early in 1920 the Poles demanded that Lenin grant Poland the 1772 boundary with Russia. (See Map 29.) Lenin refused the demands but attempted to strike a compromise. The inevitable breakdown of negotiations unleashed a bitter, wide-ranging conflict among Poles, Russians, and Ukrainians. Polish forces swept into Ukraine, taking advantage of a downturn in Bolshevik fortunes in the civil war, overrunning nearly all of the region and capturing Kiev. A decisive victory over the Bolsheviks' White enemies permitted Red military commander-in-chief Lev Trotsky to turn his attention to the Polish forces in Ukraine. The Poles were swept out nearly as swiftly as they had entered, and, as resistance against Red military operations collapsed, the Poles were pushed out of Lithuania. Within four months of the start of the war, Russian Bolshevik forces lay at the doorstep of Warsaw.

Only Piłsudski's personal leadership and French military assistance enabled the Poles to rally against the Russians and to make a successful stand on the outskirts of their capital. The "Miracle of Warsaw" turned the tables once again in the war. Bolshevik forces were driven from before the city, and their setback soon turned into a general retreat. Russian forces were expelled from Poland into Lithuania and Ukraine, and peace negotiations were opened in Riga during October. The Treaty of Riga, in which the Poles secured a substantial part of their territorial claims in the east, was not signed until the following year (1921). The Poles won a Poland larger than any had imagined possible. Post-Riga Poland not only included the western territories delineated at Versailles (part of Polish Prussia [the "Polish Corridor"], Great Poland, Galicia, Little Poland [with Cracow], part of Silesia, and Mazovia), but it now also encompassed a large slice of western Ukraine and smaller slices of Lithuania and Belarus. In 1922 Vilnius and its surrounding district were incorporated into Poland by plebiscite.

Despite the apparent victory, Riga demonstrated the difficulty of establishing firm national borders on paper when there existed no defensible geographic features to secure them. A bitter feud erupted over Vilnius between the Poles, who viewed the Lithuanians as historical but second-class national partners, and the Lithuanians, who cherished their newly won independence from both Poland and Russia. Only the political disruption and economic exhaustion of the Bolsheviks caused by revolution, civil war, and international intervention had brought the Russians to the table at Riga. They accepted the border settlement in much the same fashion as the Romanovs had accepted the "eternal peace" treaties with Poland in the 17th century—it was necessary so that the state could stabilize and consolidate its resources for a future effort to regain "Russian" territories under Polish control. Coupled with the fact that, in the west, the Germans were dissatisfied over their border with Poland, and particularly upset over the existence of the "Polish Corridor," which separated their East Prussian territories from the main trunk of Germany, Riga merely succeeded in determining that the new Polish nationalist state had to fear continuously for virtually all of its lengthy and disputed borders.

Post–World War I Poland was in poor shape for maintaining such an arduous task. To do so, the state needed to be strongly unified to concentrate all of its human and material resources in the effort. But Poland was reconstructed out of territorial pieces that for the previous 123 years had experienced three different economic, political, and ethnic developmental processes because of the varying cultures of the three partitioning powers. So new Poland, though territorially large, was internally weak. Time was needed to solidify a unification process. But, with both Germany and Soviet Russia chafing over territorial losses to Poland, time was in short supply during the interwar period.

SWEDEN

LATVIA

Riga

SOVIET
RUSSIA

GERMANY

LITHUANIA

Kaunas

Vilnius

Polatsk

Vitsebsk

Dnieper R.

Smolensk

Gdynia

Danzig

Königsberg

GERMANY

MAZOVIA

Białystok

Minsk

BELARUS

Torun

Vistula R.

Bug R.

Poznań

Oder R.

GREAT
POLAND

Warsaw

Brest-Litovsk

Breslau

SILESIA

Łódź

Vistula R.

Katowice

Cracow

*LITTLE
POLAND*

Lublin

L'viv

GALICIA

Kiev

UKRAINE

CZECHOSLOVAKIA

Košice

RUTHENIA

BUKOVINA

Vienna

Bratislava

Danube R.

AUSTRIA

Budapest

HUNGARY

ROMANIA

- - - - - Border of Polish Nationalist Claims, Versailles, 1919

- - - - - Proposed Curzon Line

••••••• Limit of Polish Army Advance, June 1920

- - - - Limit of Soviet Red Army Advance, August 1920

Poland after the Treaty of Riga, March 1921

MILES

0   50   100   150   200

0   100   200   300

KILOMETERS

Map 44: Hungary after Trianon, 1920–1939

In 1916 the Romanians had launched a militarily futile invasion of Transylvania. Their decisive defeat effectively knocked them out of the war until its late stages in 1918, when they seized the opportunity offered by a collapsing Austria-Hungary and again invaded the principality against little real opposition. This proved the deciding factor in bringing about the Alba Iulia Assembly of Transylvanian Romanian nationalists and their declaration of the union of Transylvania to Romania. In 1920, when the war's victorious Allies sat down in Versailles's Trianon Palace to decide the fate of Hungary, they recognized a fait accompli and awarded Transylvania to Romania, comfortable in their self-assurance that the ideal of national self-determination had been served. On the ground itself, Hungarian military exhaustion (resulting from the war and the 1919 battles fought against the Slovaks and Romanians during an ill-fated Bolshevik regime led by Béla Kun) and the Romanian military presence (they had even captured Budapest in suppressing Kun) ensured that the Trianon decision was implemented.

The Treaty of Trianon, signed on 4 June 1920, officially ended World War I for Hungary. When drawing the postwar map at Trianon, the victorious peacemakers, favoring national self-determination for allied peoples, had little compunction in rewarding the Romanians for their less-than-decisive military efforts during World War I by satisfying their full claims to Transylvania and more. (See Maps 45 and 46.) Although the Romanians were particularly favored at Versailles with regard to Hungarian spoils, other "friendly" peoples were not forgotten. The regions of Croatia, Slavonia, and most of Banat were recognized as Yugoslav territories in the new Kingdom of the Serbs, Croats and Slovenes. (See Map 47.) All of Slovakia, including a large southern slice with a predominantly Magyar population, was included in the new state of Czechoslovakia. All told, Trianon stripped Hungary of almost two-thirds of its former, "historic" territories, over which it had ruled for close to a millennium.

With the territorial losses there went also over 50 percent of the country's former inhabitants, including a significant minority of Magyars. These reductions essentially created the situation that Magyar nationalists had been demanding since 1848—an ethnically homogeneous Magyar nation-state, in which over 90 percent of the population belonged to the ruling national group. Yet the nationalistic Magyars declared the cost prohibitive and the process wholly unfair. They considered the Trianon borders of Hungary artificial and imposed by force, and they became one of the most outspoken proponents of revising the Versailles peace settlements in Europe. Their nationalistic slogan, "nem, nem, soha" (no, no, never), relative to accepting the Trianon borders as final, reverberated continuously and publicly throughout the country over the decades leading up to World War II.

The passionate, sometimes close to fanatical, opposition of Magyar nationalists to the Trianon Treaty serves as a good illustration of the cultural nature of Western European nation-state nationalism, with its innate human dangers. Nationalism linked solely to common ethnicity is meaningless. A Western-style nation-state is not the political manifestation of a particular group of people alone. Nationalism and the nation-state encompass much more. They require linking a unique history that took place in a precise territory to a particular group of people. Without these two additional ingredients, there can be no nationalism or its political nation-state incarnation. Since nowhere in Europe, and especially in Eastern Europe, have particular groups of people existed over time in territories completely isolated from one another, nationalist conflicts among them are endemic. There is too much historical and territorial overlap. Problems always exist regarding where to draw nation-state borders, and why. All parties involved possess their own unique demographic, historical, and economic justifications—and only rarely do national neighbors agree with each other completely.

Thus, though Trianon Hungary ideally was as ethnically homogeneous as any nationalist could have desired, it was flawed as a true Magyar nation-state because it failed to include territories, also inhabited by Magyars, that had played continuous and important roles in the thousand years of Hungarian history in Europe. For the Hungarians, Trianon was a blatant violation of Western European nation-state political culture and a national humiliation of the first order. Territories that formed parts of Hungary for centuries had been cut away by force and incorporated into neighboring, mostly newly created, nation-states that were anything but friendly to the Magyars. The Magyar nationalists' vocal and persistent reaction against the terms of Trianon caused fear among their neighbors, who had all received large slices of former Hungarian territory by that treaty. In a series of agreements signed among themselves in 1920 and 1921, Czechoslovakia, Yugoslavia, and Romania, counting on French support, banded together in a political-military alliance bloc, termed the Little Entente, aimed at common protection against Hungary. Although the Western Great Powers took the alliance seriously for some years as a guarantor of regional stability, it had little practical results and succeeded only in further entrenching national bitterness between the two sides.

Frustrated in their revisionist claims, hemmed in on all sides with avowed national enemies, and with the state economically weak because the territorial losses took traditional agricultural markets and industrial resources with them, the Magyar nationalists slipped further to the political right during the 1930s, spawning the fascist Arrow Cross movement. Although the government, dominated by Admiral Miklós Horthy, did not officially tie itself to the extremist developments, its nationalist program eventually led to a close alliance with Adolf Hitler's radically revisionist Nazi Germany.

CZECHOSLOVAKIA

BOHEMIA

• Prague

• Brno
MORAVIA

SLOVAKIA

POLAND

• Cracow

• L'viv

GALICIA

RUTHENIA

UKRAINE

• Linz
Danube R.

AUSTRIA

• Vienna

• Graz

• Klagenfurt

SLOVENIA

ITALY

ISTRIA

• Ljubljana

• Rijeka

CROATIA

• Zagreb

• Bratislava

• Sopron

• Budapest

HUNGARY

• Pécs

Košice

• Miskolc

Tisza R.

• Debrecen

• Oradea

• Cluj

TRANSYLVANIA

MOLDAVIA

• Szeged

• Arad
Murez R.

• Alba Iulia

• Timișoara

SLAVONIA

VOJVODINA

BANAT

• Sibiu

• Brașov

ROMANIA

YUGOSLAVIA

BOSNIA-
HERCEGOVINA

• Belgrade
Danube R.

SERBIA

WALLACHIA

Trianon Hungary, 1920

Pre-1918 Kingdom of Hungary

– – – Limit of Hungarian Bolshevik Military Operations, 1919

MILES

0   50   100

0   50 100 150

KILOMETERS

Map 45: Romania after Trianon, 1920–1938

Throughout the 18th century the thrones of the Wallachian and Moldavian principalities were sold by the Ottomans to members of wealthy Greek Phanariote families, who ruled their domains in Byzantine autocratic fashion, surrounded by hollow ritual and dependent landowners, and subject only to the whims of the Turkish sultans. Under Phanariote rule, one of the most oppressive and inequitable social systems in Eastern Europe emerged, with a small privileged elite of powerful magnates lording it over the majority of poor, enserfed peasants and numerically inconsequential urban inhabitants. During the 19th century the repeated Russian presence north of the Danube virtually led to the establishment of a Russian protectorate, essentially replacing Greek with Russian control in 1829.

Romanian nationalism originated among Transylvanian Romanians in the Habsburg Empire and filtered over the Carpathians into the principalities by means of economic and cultural (Orthodox) links. It exploded in the revolutions of 1848. The Principality Romanians, under heavy French influence, exhibited a highly aristocratic and essentially feudal form of nationalism that was concerned with blotting out the Greek Phanariote legacy of their Ottoman past and looked to build a Greater Romania, while the Transylvanian Romanians, whose nationalism was based on a policy of social reform for an essentially egalitarian society, struggled to gain recognized parity with other nationalities and social reform within the Habsburg Empire.

The Romanian revolution of 1848 resulted in successive periods of Russian and Austrian occupation of the Principalities until 1857. At a European Great Powers conference in Paris (1858), Wallachia and Moldavia were permitted to establish common institutions under separate princes. Alexander Cuza (1859-66) managed to win election as prince in both principalities, and, by 1862, the union was recognized as a single country, Romania. Cuza squelched all internal opposition by strengthening the legal powers of the princely office, but was kidnapped and forced to abdicate in 1866. His replacement was a German, Prince Karl I of Hohenzollern-Sigmaringen (1866-1914), who won full independence from Russia in 1878 as a reward for Romanian participation in the 1877-78 Russo-Turkish War, though at the expense of ceding Bessarabia to Russia. In 1881 Karl raised himself to the position of king. His reign was characterized by disturbances springing from the inequitable social division between the wealthy landowners and the poor peasants, which resulted in long periods of martial law.

Karl I started agitation among the Transylvanian Romanian nationalists for union with his kingdom before the outbreak of World War I. By joining the Entente in 1883, he hoped to gain Transylvania in return. Thereafter, he began making anti-Magyar public attacks aimed at gradually separating Transylvania from Hungary through the efforts of its Romanian nationalists. Although the Magyar government did its best to suppress Karl's nationalist agitation, Habsburg crown prince Francis Ferdinand, who despised and distrusted the Magyars, secretly promised to Karl the cession of Transylvania to Romania if Karl agreed to join his country to the Habsburg Empire. By the time war erupted in 1914, Karl's agitation among the Transylvanian Romanians was highly developed, but largely unsuccessful.

The collapse of Austria-Hungary in World War I led to the union of Transylvania to Romania, including large Magyar and German minority populations. (See Maps 42 and 44.) In late 1918, taking advantage of the Russian Revolution, Romania also annexed Bessarabia, which had a mixed population of Romanians, Turks, Bulgarians, Ukrainians, and Russians. The Treaty of Trianon validated both acquisitions. The Romanians then implemented programs that would weaken the minorities in their newly enlarged state and strengthen themselves. Non-Romanian governing institutions were eliminated and minority officials were systematically weeded out of their posts. Public schools became tools for Romanianizing the minorities, while minority church and private schools were either seized by the government or closed on the most specious of pretexts. Beatings and imprisonments of non-Romanians became commonplace.

Much of the rabid anti-minority policy took place behind a screen of liberal legalism. The 1923 constitution was a model of liberal-democratic ideals. The law was good; its enforcement was not. Romanian officials bullied minorities living in their districts and showed utter contempt for the laws they were sworn to uphold. The Agrarian Reform of 1920 also was an outwardly progressive policy that was used to discriminate against non-Romanians. While certainly needed in the former principalities, where traditional feudal relationships still existed, it also offered an excellent means for depriving minorities of their land, which supported their religious and educational activities, thus weakening the minorities' national aspirations and giving the government direct control of minority education. The minority problem was also aggravated by rampant anti-Semitism among the peasantry, who viewed the influx of Jewish land stewards, who were refugees from Poland and its Catholic-charged nationalist euphoria, as another form of landowner oppression.

When royal power declined under King Ferdinand (1914-27), the facade of liberal democracy proved unable to exert stable government because of social unrest over land reform, anti-Semitism, and minorities. Political assassinations became the norm in Romanian politics. King Karl II (1930-45) was forced to deal harshly with the "Iron Guard" (officially the "Legion of the Archangel Michael"), one of Europe's earliest fascist movements, by establishing a royal dictatorship and ordering the murder of the Guard's leaders in 1938.

CZECHOSLOVAKIA

POLAND

UKRAINE

Dniester R.

Southern Bug R.

Chernivtsi ● ● Hotyn

BUKOVINA

BESSARABIA

Dniester R.

Satu-Mare ●
MARAMUREŞ

Tisza R.

Debrecen ●

HUNGARY

Oradea ●

Someş R.

Prut R.

Iaşi ●

CRISANA

Cluj ●

Tîrgu-Mureş ●

MOLDAVIA

Kishinev ●

Odessa ●

Szeged ●

Alba Iulia ●

TRANSYLVANIA

Arad ●

Mureş R.

Sibiu ●

Olt R.

Braşov ●

Galaţi ●

Izmaïl ●

Timişoara ●

BANAT

Danube R.

Brăila ●

WALLACHIA

Sava R. Belgrade ●

Turnu-Severin ●

Ploieşti ●

Argeş R.

Bucharest ●

DOBRUDZHA

BLACK
SEA

YUGOSLAVIA

Craiova ●

Giurgiu ●

Constanţa ●

Vidin ●

Danube R.

Ruse ●

Silistra ●

BULGARIA

Varna ●

MILES

0	50	100

Pre-1918 Romania

Romania's Trianon acquisitions, 1920

Border of Romanian Nationalist Claims, Versailles, 1919

0	50	100	150

KILOMETERS

Map 46: The Transylvanian Question

While Magyar nationalists ardently sought to reclaim all the territories and populations torn from "historic" Hungary by Trianon, no loss was more galling or more fervently disputed than Transylvania's. (See Map 44.) Their attempts to gain international redress on this issue began at Versailles in the negotiations leading up to the treaty. They were actively continued through the Minorities Question Section of the League of Nations' Secretariat and through the national and international print media. The Magyar nationalists' unwillingness to let their Transylvanian cause subside from international notice forced the Romanian nationalists to respond in kind. Although from 1919 until 1940 both the Magyars and the Romanians poured mountains of statistical data relating to demographic, economic, and political issues into supporting their respective arguments, the heart of both sides' cases justifying their conflicting claims on Transylvania was historical. The incessant public dispute between them soon became known as the "Transylvanian Question." It lasted as an open diplomatic and media sore until 1940, when Hitler attempted to force a compromise solution on the two sides; thereafter, the end of World War II and the subsequent submergence of both contending parties beneath the tide of communism hid the conflict below the surface of Soviet-imposed international socialist brotherhood.

Transylvania had been incorporated into the medieval Hungarian state in the 11th century under King St. István I. (See Map 13.) During the 16th and 17th centuries it had enjoyed a golden age of quasi-independence as a Hungarian bastion of anti-Habsburg Protestantism (see Map 24), and in the Hungarian Revolution of 1848-49 it had served as the final fortress of the revolutionaries against the forces of both the Habsburgs and the Russian Romanovs (see Map 33). To Hungarian nationalist minds, it was unthinkable that a region of such historical national importance should be placed into the hands of Romanians. Through continuous representations before the League of Nations and its minority rights commission in Geneva, voluminous publications in Western languages, and constant political agitation both within and outside of Hungary between 1920 and 1939, the nationalist Hungarians kept the Transylvanian issue burning on the stage of international public opinion. The Hungarians agitated so rabidly for a revision of Trianon that their neighbors, all of whom had received slices of Hungarian territory in the treaty, formed the Little Entente, to protect themselves from possible Hungarian efforts to revise Trianon by force. (See Map 44.) This only inflamed Hungarian nationalism further.

Romania was forced to publicly counter the Hungarian historical arguments for control of Transylvania. The Hungarians claimed that Transylvania was uninhabited when their ancestors conquered the region in the 11th century. For that reason they had been forced to bring Hungarians from Pannonia to settle there, as well as to invite in outside colonists—Székelys (ethnic relatives of the Hungarians) and Saxon Germans—to help protect and economically develop the region. According to the Hungarians, Romanians only entered Transylvania in any numbers (from the Balkans to the south) starting in the late 12th and early 13th centuries. Since the newcomers were a lowly peasant people, the Romanians remained of little political or cultural account compared to the original Hungarian, Székely, and Saxon populations, who shouldered the burdens of administering, protecting, and developing the region.

The Romanians claimed the region by right of demographic majority—55 percent of the region's population at the time of Trianon was Romanian—and by right of historical possession. According to their case, the Romanians' ancestors were the ancient Daks, who controlled the region of Transylvania and who were conquered by the Roman Emperor Trajan (98-117) and placed under Roman authority for 160 years, during which time they were Latinized. When Rome withdrew from the region in 271, the Romanized Daks remained behind, finding refuge in the mountains from the tidal waves of Germanic and Asiatic migrations that swept through the region over the course of the 3rd through 9th centuries. They would reemerge in the Transylvanian lowlands by the 11th century, when they were conquered by Hungarian intruders. The Romanians' theory of their origins thus reduced all the other nationalities in Transylvania to historical interlopers in a native Romanian homeland.

In 1939 both Hungary and Romania found themselves allied with Germany—the Hungarians because of a sense of nationalist-revisionist common cause with the Germans and the Romanians out of economic necessity. Hitler needed both —the Hungarians to ensure his political dominance in Eastern Europe and to protect his southern flank and the Romanians to provide necessary oil and manpower resources— for his planned future military operations in the East. In 1940, when the conflict between Hungary and Romania over Transylvania threatened to explode into warfare, Hitler attempted to impose a solution to the problem. In the Second Vienna Award (30 August 1940) Hitler gave the northern 40 percent of Transylvania, including the Székely region in the extreme southeast, with a population of some 2.5 million people (52 percent of whom were Magyars), to Hungary.

Although Hitler considered the matter settled, and the two protagonists had no choice but to accept his dictate, the Vienna solution solved nothing. Neither Hungarians nor Romanians considered the award anything other than a temporary settlement that would be worked out once Germany won World War II. Its immediate result was the disaffection of both allies. Unfortunately for the Hungarians, the award was nullified by German defeat. Romania defected at the last moment to the anti-German allies at the end of 1944 and received all of Transylvania—and more—as a reward.

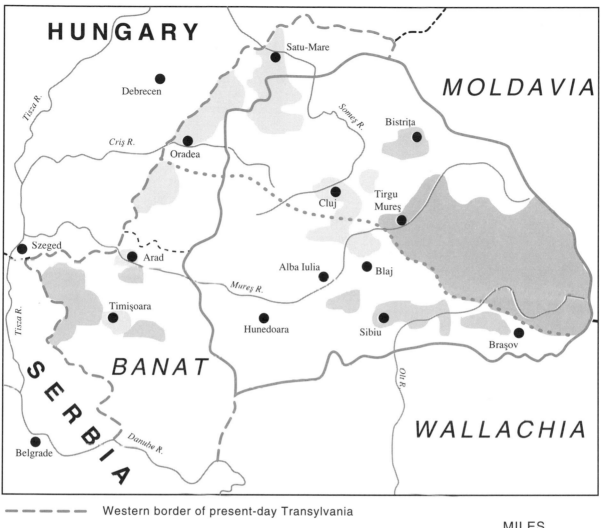

HUNGARY

Tisza R.

Debrecen

Cris R.

Oradea

Satu-Mare

Somes R.

Bistriţa

Cluj

Tirgu
Mureş

MOLDAVIA

Szeged

Arad

Mures R.

Alba Iulia

Blaj

Tisza R.

Timişoara

Hunedoara

Sibiu

Olt R.

Braşov

SERBIA

BANAT

WALLACHIA

Belgrade

Danube R.

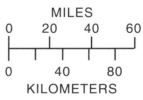

- - - - - Western border of present-day Transylvania

———— Border of "historic" Transylvania

· · · · · · · · Southern limit of 1940 Second Vienna Award

Areas with historical Hungarian ethnic majority

Areas with historical Székely ethnic majority

Areas with historical Saxon ethnic majority

MILES

0   20   40   60

0   40   80

KILOMETERS

Map 47: Versailles-Created Yugoslavia, 1921–1941

In July 1917 Ante Trumbić, Croat leader of a Serb-leaning, Serbo-Croatian faction from Dalmatia, and Serb prime minister and nationalist, Nikola Pašić, met on Corfu and signed a pact calling for the establishment of a Yugoslav state following World War I. Pašić agreed to Croat demands for a constitutional monarchy responsible to a democratically elected national assembly as the governing framework for the state. He did so, however, only because of political circumstances: the Serbs lacked the usual Russian support for their Greater Serbia claims (the revolution had overthrown the tsar); the Americans, newly involved in the war, favored the Croat-inspired federalist Yugoslav idea; and Serbia was occupied by enemy Austrian and Bulgarian forces.

Despite Pašić's apparent compromise in the agreement, the Serbs continued their ultranationalist Greater Serbia Yugoslav approach for the duration of the war, much to the apprehension of the Croat and Slovene Yugoslav federalists. As the fighting ended in 1918 and the Habsburg Empire disintegrated, a series of declarations proclaiming South Slav independence by the various "Yugoslav" nationalists of Serbia, Croatia, Montenegro, and Slovenia culminated in a final joint pronouncement in December 1918 from Belgrade formally establishing the Kingdom of the Serbs, Croats and Slovenes. The United States was the first Great Power to recognize the new state in February 1919, and the victorious Entente mapmakers at Versailles swiftly followed suit by stamping their approval into the treaties produced.

Doubts as to the sincerity of Serb promises to accommodate their new national partners surfaced quickly. In a 1920 plebiscite approximately half of the Slovenes, most of whom were Germanized to the point of being "almost German," chose to remain with Austria rather than join the new South-Slav kingdom. The Croats, and those Slovenes who had decided not to stay with Austria, soon found that concerns over ties to the Serbs were valid. The Serbs soon organized the state as a centralized Serb nation-state under strong royal control. Both the Croats and the Slovenes envisioned the political structure of the state in more liberal-democratic and federalist terms that would grant them significant local autonomy. Neither Croats nor Slovenes were happy with the thought that the Serbs—who to their minds were cultural inferiors whose royal house stemmed from glorified, illiterate pig farmers a century earlier—intended to play the role of dominant partner in a highly centralized state. While the Slovenes were able to maintain somewhat of a low profile in terms of political opposition to the situation because Croatia separated them from direct contact with the Serbs, the Croats, by their very geographical position next to Serbia, were forced to play the leading oppositional role.

Croat nationalist opposition exploded into extremism under Serb king Aleksandr I Karadjordjević (1921-34), who was particularly intent on consolidating the state under his central authority. In 1924, when Croat boycotts of elections and parliamentary proceedings initially failed to win political concessions from the Serbs, the Croats, led by Stjepan Radić, eventually entered the political process and won seats in both the parliament and the royal cabinet. But after Radić was fatally shot in the parliament's chamber by a radical Serb in 1928, the Croats withdrew their political participation in Belgrade, demanded a new federal constitution, and established their own separatist government in Zagreb, Croatia's capital. In response the next year, Aleksandr proclaimed a royal dictatorship, dissolved the nationalist Croat political party, and arrested Croat leaders. The country was renamed Yugoslavia and administratively reorganized on a purely geographical basis in an effort to eliminate all traces of historical-national associations. A new constitution was issued in 1931, which rigged any political participation in favor of the Serb-dominated royal government, and Croat and Slovene oppositional leaders continued to be arrested.

Out of this unbalanced situation there arose among the Croats an extremely radical, ultranationalist terrorist organization—the *Ustaše*. It forged close relations with outside Hungarian and Italian fascist leaders and internal Bulgaro-Macedonian terrorists. In 1934, from its headquarters in Hungary, the *Ustaše* masterminded the assassination of King Aleksandr in Marseilles. Less radical Croats offered to cooperate with Prince Pavel, the regent of the dead king's young successor, Petr II (1934-41), but he reneged on his promised political concessions to them, thus tilting Croat opposition back once again toward the extremists. This trend was reinforced in the more strictly cultural sphere when, in 1937, Pavel's regency government dropped a project that, in accordance with a concordat made with the Vatican, would have granted wider privileges to Roman Catholics (that is, Croats and Slovenes) following widespread disturbances instigated by Orthodox groups and Serb radicals who opposed any such arrangement.

Rising Croat opposition, reinforced by a growing democratic movement among more farsighted Serbs, led to serious discussions with the royal government in 1938 and 1939 regarding conciliatory moves to relieve the pressures building in the country. Royal dictatorship was ended and plans were laid for federalizing Yugoslavia, with the Croats to receive full cultural and economic rights. Croat leaders were readmitted to the royal cabinet. In March 1941 this progress came to an end when a successful military coup against Pavel and his pro-German policies placed the youthful Petr on the throne. Ten days later Hitler invaded Yugoslavia, with the help of Italian, Hungarian, Bulgarian, and Romanian allies, and crushed the Yugoslav forces in less than two weeks. He then established a friendly Croat *Ustaše* government to hold most of the country while he went about prosecuting his invasion of the Soviet Union. (See Map 49.)

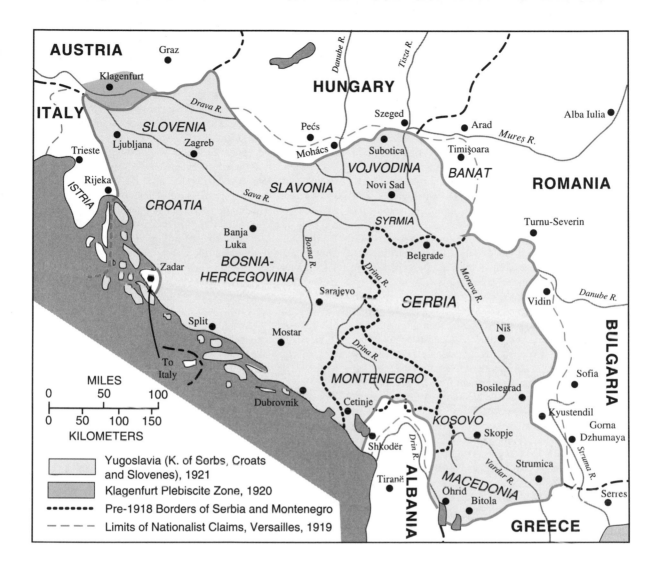

AUSTRIA

Graz

Klagenfurt

ITALY

Trieste

Rijeka

ISTRIA

Zadar

Split

Dubrovnik

HUNGARY

Drava R.

SLOVENIA

Ljubljana

Zagreb

Pećs

Mohács

Danube R.

Tisza R.

Szeged

Subotica

VOJVODINA

Novi Sad

Arad

Timişoara

BANAT

Mureș R.

Alba Iulia

ROMANIA

SLAVONIA

Sava R.

CROATIA

Banja
Luka

BOSNIA-
HERCEGOVINA

Bosna R.

SYRMIA

Belgrade

Drina R.

Sarajevo

SERBIA

Morava R.

Turnu-Severin

Vidin

Danube R.

BULGARIA

Niš

MILES

0   50   100

0   50   100   150

KILOMETERS

To
Italy

Mostar

MONTENEGRO

Cetinje

Drina R.

Shkodër

Drin R.

Sofia

Bosilegrad

Kyustendil

Gorna
Dzhumaya

KOSOVO

Skopje

Tiranë

ALBANIA

MACEDONIA

Ohrid

Bitola

Vardar R.

Strumica

Struma R.

Serres

GREECE

	Yugoslavia (K. of Sorbs, Croats and Slovenes), 1921
	Klagenfurt Plebiscite Zone, 1920
•••••	Pre-1918 Borders of Serbia and Montenegro
––––	Limits of Nationalist Claims, Versailles, 1919

Map 48: Czechoslovakia and Munich, 1920–1939

Before Versailles, Czechoslovakia had never existed. It possessed no historical precedents or traditions. The new state's borders joined together Bohemia-Moravia and Slovakia, which had been separate since at least the 10th century. It was primarily the creation of Czech (Tomáš Masaryk and Edvard Beneš) and Slovak (Milan Štefánik) World War I emigré nationalists and their followers in France and the United States. Masaryk's Pittsburgh Agreement (1918) with American Slovak emigrés, which earned him the first presidency in the new country, formed the basis for the new state. Although Czechoslovakia's constitution was the most liberal-democratic of all the Eastern European states that emerged after World War I, and it quickly won a reputation as a model of functioning democracy, the state was fatally flawed in ethnonational terms.

Czechoslovakia was created essentially on the basis of a hyphen—two separate parts and peoples joined together to shape one, supposedly unified, state. Soon after Czecho-Slovakia became a reality, Slovak nationalists began insisting that they had been duped. Czech nationalists denied those claims, but political realities tended to confirm the Slovaks' awkward position in the arrangement from the start.

It was impossible to create a unitary state consisting of two equal halves out of halves that were unequal. The Czechs possessed a long history in the political life of the Habsburg Empire. Prague, the capital, had been a cultural and commercial center for Central-Eastern Europe since the 14th century. It was the home of Panslavism, the earliest expression of modern nationalist thought among both Czechs and Slovaks. Under the Habsburgs, the Czechs had maintained a voice in the imperial diet in Vienna. On the other hand, the Slovaks, who had existed under Hungarian rule for close to a thousand years, had played no political role in the Hungarian portion of the old Habsburg Empire and had been virtually voiceless in the Budapest diet. They became highly Magyarized over time because that was the only avenue open to them for advancement.

The inequality was obvious in numerous areas. While Czech Bohemia was heavily industrialized, urbanized, and liberal-minded, mountainous Slovakia remained agriculturally primitive, backward, and conservative. Czechs were nominally Catholics but were not particularly devout. Slovaks were strongly Catholic. Czechs tended to be secular, materialistic, anticlerical, and open to socialism. (In the interwar years, the percentage of socialists in Czechoslovakia was higher than in any other European country.) Slovaks were more susceptible to leadership by their clergy, who traditionally filled the ranks of the Slovak nationalists. Inevitably, the Czechs played the leading role in government, economics, and education in the state. This could not help but create suspicion in both ethnonational parties, with most of the suspicion stemming from the lesser partner, the Slovaks. By the time Hitler set out to obliterate Czechoslovakia in 1938, thus laying open Central-Eastern Europe to German domination, the unbalanced Czech-Slovak relationship had led to the rise of a nationalist Slovak separatist movement. Soon after England and France sold out to Hitler at Munich, an independent Slovakia was proclaimed, but its independence was based on Nazi German support, and it lasted only as long as did its fascist ally.

The pre-Munich Czech-Slovak relationship was exacerbated by other internal tensions. A third ethnonational group had been included within Czechoslovakia—the Ruthenians. All the unequal comparisons regarding the position of the Slovaks relative to that of the Czechs were true for them. They possessed no historical connections to the Czechs whatsoever. They, like the Slovaks, had been long associated with the Hungarians, and their territory was even more mountainous and backward. Their Uniate-Catholic church was controlled by Slovaks and provided the basis for the Ruthenians' nascent sense of nationality, since it cut them off from ethnically related Ukrainians. It also laid them open to Slovak cultural-nationalist claims that they were Slovaks. Faced by Czech dominance and Slovak nationalist claims, the Ruthenians were virtual nonentities in Czechoslovakia. They were forcibly incorporated into Hungary in 1939.

Less welcome but far more troublesome than the Ruthenians were the Hungarian and German minorities. Numerous Magyars inhabiting a broad swath of southern Slovakia never accepted their position in Czechoslovakia and expressed a strong sense of Magyar nationalism. When Hitler set about dismembering Czechoslovakia following the Munich sellout, the Hungarians of Slovakia proved his ready accomplices. In the First Vienna Award (1938), Hitler gave the Hungarian-inhabited area of Slovakia to Hungary, which held it only until the close of World War II.

Even more instrumental in Hitler's destruction of Czechoslovakia at Munich were the Germans in Bohemia, who inhabited a long belt of territory in the mountainous Sudet, Ore, and Bohemian Forest border regions, which collectively came to be known as the Sudetenland. Directly across the Versailles-imposed frontier lay defeated and nationalistically frustrated Germany. Although the Masaryk-Beneš Czechoslovak government made honest attempts to integrate their economically important German minority into the new state, the Sudeten Germans eventually succumbed to the frenzied hysteria of Nazi German ultranationalism and became willing tools of Hitler in his efforts to destroy the country. That was accomplished in 1939, when Bohemia-Moravia was annexed by Germany.

Joining in the dismemberment of Czechoslovakia after Munich was Poland. There a long-standing feud had existed between the two states over possession of the rich region around Téšín (Cieszyn). Hitler handed it to the Poles (1938).

1	Sudetenland. Occupied by Germany after Munich, September 1938
2	Cieszyn. Occupied by Poland after Munich, September 1938.
3	Occupied by Hungary under First Vienna Award, November 1938.
4	Occupied by Hungary, March 1939.

———— Borders of Czechoslovakia, 1920

– – – – – Limits of Nationalist Claims, Versailles, 1919

•••••••• Border of Czechoslovakia after Munich (September 1938 to March 1939)

MILES
0   50   100

0   50   100   150
KILOMETERS

Map 49: Eastern Europe during World War II, 1938–1944

The destruction of Czechoslovakia freed Hitler to begin his war of German conquest in Eastern Europe by invading Poland in September 1939. He had prepared for the war carefully by taking advantage of national, political, and economic conditions among various East European states. Both Hungary and Bulgaria were loser states at Versailles and strongly revisionist. Ultranationalist Hungary had strong ties to Hitler's fascist ally Italy dating to 1927 and a growing fascist movement of its own, the Arrow Cross. Bulgarian king Boris III (1918-43) was ethnically German and personally sympathetic to the anti-Versailles lead taken by Germany. Romania, a Versailles winner, was wracked by social problems that spawned rabid anti-Semitism and an early fascist movement, the Iron Guard. In Yugoslavia, another Versailles winner, a royal dictatorship, under the regency of pro-German Prince Pavel, desperately tried to curb Croat national animosities against Serb political domination.

All of those states, essentially agrarian, were hard hit economically by the Great Depression. Hitler offered them lucrative trade agreements essentially on a barter basis — German industrial and manufactured goods in return for basic food products and raw materials — that tied them dependently to Germany. He then sweetened the pie for Hungary, which he considered crucial for his future plans since it could provide him with manpower and serve as a Damocles' sword in keeping neighboring states in line with his plans, by granting it large slices of Czechoslovakia.

In August 1939 German foreign minister Joachim von Ribbentrop and Soviet foreign minister Vyacheslav M. Molotov signed a nonaggression agreement known as the Ribbentrop-Molotov Pact. It contained a secret protocol dividing Eastern Europe into mutual spheres of influence. The Soviets received the independent Baltic states (which included Lithuania), as well as Bessarabia (Moldova) from Romania, and the right to annex those portions of Belorussia and western Ukraine east of Versailles' Curzon Line that Poland had attained at Riga in 1921. (See Map 43.) Germany was given a free hand in the rest of Poland and Central-Eastern Europe. By this pact, Hitler accomplished steps necessary for his invasion and conquest of Poland.

By using the existence of the Polish Corridor, a strip of Polish-held land cutting off German East Prussia from Germany proper, created by the Versailles Treaty to give Poland direct access to the Baltic Sea, as a pretext for initiating an invasion of Poland in 1939, Hitler sparked World War II. Exposed to overwhelming German military forces to their west and north, and then invaded by the Soviets, who were seeking the lands granted them by the Ribbentrop-Molotov Pact, the Poles were swiftly overcome. The Nazis then set in motion a propaganda campaign inside Germany portraying Slavic Poles as somewhat subhuman, which lent perverted credence to the genocidal policies they followed inside conquered Poland during the four years of occupation. At least 3 million Poles were exterminated during World War II, and those who escaped that fate were subjected to the most degrading conditions. The Nazis transformed Poland into the chief killing ground in their efforts to rid their future Europe of those they considered undesirable, human "vermin" — Jews, Eastern Slavs, Gypsies, political opponents, among others — through methodical extermination.

As German victories in the north and west multiplied in the first year of the war (1940), Hungary officially was brought into the Axis alliance and, in the Second Vienna Award, granted a large slice of Transylvania at Romanian expense. (See Map 46.) Also, Romania was forced to cede southern Dobrudzha to Bulgaria (Treaty of Craiova). The Romanians were partially mollified by Hitler's guarantee of the remaining borders of their state and full participation in the Axis alliance. These moves were efforts to counter the Soviet Union's determination to expand its influence in the Balkans, especially in Bulgaria and Romania. Hitler could ill afford that development, since Romania was a source of much-needed petroleum supplies, which Germany lacked, and a potential source for manpower, while a friendly Bulgaria assured Germany's southern flank against possible Soviet encroachment in the eastern Mediterranean. Stalin's refusal to relinquish Soviet interests in the Balkans helped Hitler decide on war against the Soviet Union in 1941.

German plans were upset when Mussolini's Italy unsuccessfully attacked Greece in late 1940 to gain control of its Adriatic coast and found itself desperate to hold onto its protectorate, Albania, against Greek attacks in 1941. To protect his flank, Hitler brought Bulgaria into the Axis and shipped troops to that state. When Pavel of Yugoslavia was cowed into joining the Axis also, the Serbs rose against him, instated King Petr, and attempted to win Hitler's acceptance. Instead, he invaded Yugoslavia and Greece simultaneously (April 1941) and crushed both in less than a month. Greece was occupied by Germany. Yugoslavia was broken into an *Ustaše*-controlled, allied fascist Croatia and a German-occupied Serbia. Hungary received part of Vojvodina, Bulgaria received most of Macedonia and a slice of Serbia, and Italy acquired Montenegro, with parts of Serbia and Macedonia. Large-scale anti-Serb atrocities soon erupted, resulting in the rise of Communist (led by Jozip Tito) and Serb *četnik* (led by Draža Mihajlović) partisan activities that continued until the Balkans were liberated from German control in 1944.

Soon after finishing his Balkan Campaign, Hitler invaded the Soviet Union. In return for its petroleum and massive troop contributions, Romania, now heavily under fascist Iron Guard influence, regained Bessarabia and was granted an additional strip of territory, known as Transdnistria, which included the important Black Sea port of Odessa.

SMOLENSK

LITHUANIA
Vilnius
Danzig
Königsberg
GERMANY
(Part)
Minsk
BELARUS
Berlin
GERMANY
Warsaw
POLAND
SOVIET
UNION
Munich
Danube R.
Prague
BOHEMIA-MORAVIA
Cracow
Lublin
L'viv
Kiev
SLOVAKIA
GALICIA
UKRAINE
Vienna
Bratislava
AUSTRIA
Budapest
Debrecen
HUNGARY
Iaşi
BESSARABIA
TRANSDNISTRIA
Odessa
Venice
Ljubljana
Rijeka
Zagreb
Szeged
Timişoara
Cluj
TRANSYLVANIA
Sibiu
Braşov
VOJVODINA
BANAT
ROMANIA
Zadar
CROATIA
Sarajevo
BOSNIA-
HERCEGOVINA
Belgrade
SERBIA
Bucharest
DOBRUDZHA
Constanţa
To
Italy
Split
Danube R.
Ruse
ITALY
Rome
Dubrovnik
MONTENEGRO
Cetinje
Sofia
Varna
BULGARIA
Naples
ALBANIA
Skopje
Tiranë
To
Italy
MACEDONIA
Kavalla
Istanbul
Thessaloniki
GREECE
Ioannina
TURKEY
(Neutral)
SICILY
Athens

MILES
0 50 100 150 200
0   100   200   300
KILOMETERS

DODECANESE
IS.
To
Italy
CRETE

- · - · - 1938 Borders
- ·· - ·· - 1943 Borders
———— Russo-German Border Agreements, Aug. 1939-June 1941
GREECE States Annexed or Occupied by Axis Powers

Map 50: Eastern Europe, 1948–1991

In 1943 Hitler's Nazi Germany and German-dominated Eastern Europe began to collapse on the eastern front with massive military defeats against the Soviet Union at Stalingrad and Kursk. Disintegration accelerated throughout 1944, when the Red Army pushed German forces out of Russia into Poland, a western front opened against Germany after the Normandy invasion, and a relentless Western air campaign on Germany caused major communications and logistical problems and widespread disillusionment among the Germans at home. Soviet Red Army forces entered the Balkans in August 1944, leading to the surrender of Romania in that month, the collapse of Bulgaria in the next, and a link up with Tito's now dominant Communist Yugoslav partisans by October. Outflanked, Germany abandoned its remaining occupied states, took control of Hungary, and made a last stand in Eastern Europe at Budapest. To the north, the Soviets drove deeper into Poland, ceasing temporarily at Warsaw (August) long enough to permit the Germans to crush a Soviet-encouraged uprising of Polish nationalist forces before pushing farther west.

The Yalta Conference (February 1945) among Josef Stalin, Franklin D. Roosevelt, and Winston Churchill set the tentative groundwork for post–World War II Eastern Europe. Facing the military reality of the Red Army presence in most of Eastern Europe, Stalin won acceptance for a strong Soviet "influence" in return for promises guaranteeing "free and unfettered" postwar elections in those states. The prewar boundaries of the East European states were reconfirmed, except for those of Poland, which was to be stripped of its Ukrainian and Belorussian territories east of the Versailles Curzon Line. (See Map 43.) At the later Potsdam Conference (July-August 1945), Stalin was constrained to agree to a readjustment of Polish borders — its western boundary was pushed farther into traditional German territory to compensate for lands lost to the Soviets in the east.

The elections promised by Stalin at Yalta never materialized for the most part. In its train, the Red Army brought to those states liberated from the Germans Soviet-trained native Communists, who quickly began to exert Soviet "influence" through Popular Front governments. Ostensibly coalitions of Communist and traditional parties, with the Communists often holding numerically small numbers of ministry seats, the Fronts proved mere tools for the consolidation of Communist control, since the seats held were usually the most crucial (interior, foreign affairs, treasury, etc.). By 1948, capped by the "suicide" of Czech president Jan Masaryk in Prague, the Communists emerged as the holders of power in every East European state, including Soviet-occupied East Germany, except Greece, where a British war presence (1944) and then American intervention in a Communist-spawned civil war (1946-49) helped prevent that fate. The Soviets then tied all these states closely to Soviet interests by creating the Council for Mutual Economic Assistance (COMECON), which integrated the economies of its members with that of the Soviet Union.

The liberal-democratic, capitalist Western states, led by the United States, countered the creation of a Communist cordon sanitaire for the Soviet Union in Eastern Europe with first the Marshall Plan (1947) and then the Western European military alliance, the North Atlantic Treaty Organization (NATO) (1949), which both Greece and Turkey joined. The Soviets responded in 1955 by founding the military Warsaw Treaty Organization (Warsaw Pact), which included all Communist states in Eastern Europe except Yugoslavia (later, in 1960, Albania withdrew).

The Yugoslav exception was significant. Tito, a hero to the anti-fascist elements in Yugoslavia and head of state, soon came to resent Stalin's efforts to micromanage Yugoslav developments. He broke with Stalin in 1948 and went on to fashion a sort of national communism that permitted experimentations with mixed socialist-capitalist economics while retaining the iron hand of overall Communist political authority. Only he proved capable of holding the state together in the face of competing historical ethnonational stresses. His death (1980) led to a decade of internal political collapse and economic decline, culminating in the violent disintegration of Yugoslavia in 1991. Albania, originally subordinate to Tito's organization, used the 1948 split to assert its independence by supporting Stalin. But when Nikita Khrushchev repudiated Stalin and Stalinism in 1960, Albania broke with the Soviet Union and the Warsaw Pact, aligning itself first with Maoist China and then, following China's rejection of strict Maoism, attempted to go it alone.

Over the course of the cold war years, some of the East European Communist states attempted to follow Yugoslavia's example by initiating policies that smacked of national communism in various degrees, but with fateful results. Poland (1956), Hungary (1956), and Czechoslovakia (1968) learned that, so long as the Soviet Union continued to consider a unified bloc of subservient East European states a necessity for its security, deviation from Soviet-controlled Warsaw Pact and COMECON policies would not be tolerated. All three suffered military invasion to force them back into line.

Fatal flaws in dogmatic Marxist ideology, and a growing exposure to Western capitalist materialism stemming from thaws in the cold war during the 1970s and 1980s, led to increasing disillusionment with the Communist system in general. Mihail Gorbachev's attempts to counter both of those issues in the late 1980s by his *perestroika* and *glasnost* reforms of Soviet communism failed in their primary goals. Both policies demanded that the Soviets relinquish their direct hold over Communist Eastern Europe. As soon as that became a reality, the forces unleashed led not to reform of Eastern European communism but to its outright rejection.

WEST GERMANY

To Russia

LATVIA

RUSSIA

LITHUANIA
Vilnius
Kaliningrad
Gdańsk

Smolensk

Minsk

BELARUS

Berlin

EAST
GERMANY

Leipzig

Oder R.

Poznań

Warsaw

POLAND

Neisse R.

Dresden

Wrocław

Prague

CZECHOSLOVAKIA

Cracow

L'viv

U. S. S. R.

Kiev

UKRAINE

CZECH
REP.

Danube R.

Brno

Munich

Košice

Bratislava

Vienna

SLOVAKIA

Debrecen

AUSTRIA

Budapest

HUNGARY

MOLDOVA

Iaşi

Odessa

Kishinev

SLOVENIA

Ljubljana

Zagreb

Szeged

Cluj

Venice

Trieste

CROATIA

Timişoara

ROMANIA

Rijeka

BOSNIA-
HERCEGOVINA

Sarajevo

Belgrade

Zadar

Split

SERBIA

Bucharest

Constanţa

ITALY

YUGOSLAVIA

MONTENEGRO

Niş

Danube R.

Ruse

Rome

Dubrovnik

Cetinje

Sofia

Varna

Naples

ALBANIA

Skopje

BULGARIA

Tiranë

MACEDONIA

Edirne

Thessaloniki

Istanbul

Ioannina

GREECE

TURKEY

SICILY

MILES
0   50   100 150 200

0   100   200   300
KILOMETERS

Izmir

Athens

Patras

DODECANESE
IS.

CRETE

- - - - 1938 Borders
—————— Western Boundary of the Warsaw Pact States, 1955–1991
GREECE NATO States, 1949–1991
●●●●● Borders of New States, 1989–1991
LATVIA Names of New States, 1989–1991

Map 51: Wars of Yugoslav Succession, 1991-1995

Communist Yugoslavia was the federal construct of Tito, whose consolidation of power in the state depended on embracing the assorted national aspirations of prewar Yugoslavia's ethnic populations. Only Tito's ironclad central authority and determination to assert the overriding priority of Communist ideology kept the state together. His split with Stalin (see Map 50) led to the formation of Yugoslav "national communism," with the emphasis on "Yugoslav" (rather than on Yugoslavia's various constituent nationalities).

Titoist federalism intended to satisfy the prewar nationalist ambitions of Yugoslavia's diverse population within the overall context of a unitary Communist state by providing six Yugoslav "nationalities" with their own republics. While the Serbian, Montenegrin, Croatian, and Slovenian republics were predicated on long-standing national identities, Macedonia appeared as an official national entity for the first time in its history and Bosnia-Hercegovina, ethnically and religiously divided, was recognized as a unitary republic. Despite its federalism, however, Communist Yugoslavia was the most volatile focus of latent nationalist strife in the Communist-era Balkans.

Although federalism initially satisfied many Croatian nationalist claims, Yugoslav communism's decentralization loosed restraints on intellectual activities (among which were resurrected allusions to prewar nationalism). Intense debates between Croats and Serbs over the nature of "Yugoslavism" degenerated into old national sentiments and mutual distrust by the 1970s. For their part, the Serbs perceived national problems stemming from Serbia's autonomous, mostly non-Serb provinces of Kosovo and Vojvodina. They had been granted voting privileges equal to the republics' (Serbia's) in the central federal government (1966) and non-Serbs came to dominate their administrations. Regarding the Muslim majority in Bosnia-Hercegovina, the federal government never could decide if or how they constituted an "ethnic nationality" (as opposed to being either Croats or Serbs), while the Bosnian Muslims acquired a political structure that resembled more a *millet* than an ethnonational identity (giving them alone an awareness of Bosnia-Hercegovina as a unique political entity). When a revolving collective federal presidency among leaders of the constituent republics was instituted following Tito's death, growing national tensions led to its general ineffectiveness.

By the 1980s, Serb-Albanian nationalist tensions in the Serbian autonomous province of Kosovo broke out into sporadic open conflicts (see Map 52), opening the way for Serbian Communist strongman Slobodan Milošević to play the nationalist card in his bid for authoritarian power at the time that the Gorbachev reforms' failures were unraveling European communism. Beginning in 1988 Milošević championed the tenets of "Greater Serbia" nationalism within Yugoslavia, trampling on the autonomy of the Kosovo and Vojvodina Serbian provinces and raising the specter of Serbian domination throughout the federal state. The Croats,

Slovenes, and Macedonians, resentful of de facto Serbian predominance prior to Milošević, found the new situation unacceptable and initiated a series of political moves to disassociate themselves from Serbian control. As tensions mounted, Milošević supported Serbian ultranationalists in Croatia Proper, Slavonia, and Bosnia-Hercegovina, which ultimately brought about the succession of Slovenia and Croatia from Yugoslavia (June 1991) as independent, non-Communist national states. By year's end Bosnia-Hercegovina and Macedonia followed their lead.

Milošević sent the Yugoslav National Army (JNA) into Slovenia (June 1991) but it was stymied humiliatingly, after which it set upon Croatia in support of Serbian nationalists who rose in eastern Slavonia and Croatia Proper, where they had declared an independent region called Krajina centered on Knin. Vicious fighting between Serbs and Croats erupted around Vukovar, and the first evidence of the Serbian "ethnic cleansing" policy emerged. Despite Western European recognition, Croatia, led by the nationalist Franjo Tudjman (1991-99), was defeated and lost a quarter of its national territory.

Fighting then spilled over into Bosnia-Hercegovina, whose Muslim leader Alija Izetbegović (from 1991) unsuccessfully attempted to maintain a united state in the face of Milošević-supported Serbian nationalists led by Radovan Karadžić, who proclaimed a Serbian Republic of Bosnia-Hercegovina (March 1992) in unity with Milošević-led Serbia. In April 1992, fighting erupted between the two sides in Sarajevo and swiftly spread throughout the state. The Bosnian Serbs placed Sarajevo under siege and, with help from Milošević's JNA, gained control of two-thirds of the state by the end of 1992. A concerted "ethnic cleansing" policy was conducted in territories under Serbian occupation in an attempt to cement a permanent Serbian ethnic presence. Croatian nationalists joined in the fighting against the Muslims. Those in Hercegovina set up a Croat Herceg-Bosna republic (July 1992) and initiated their own "ethnic cleansing" activities.

The three-way war raged on through 1993 until 1995, resulting in enormous casualties, mass emigrations, and widespread destruction of private and historical property (including the historic Ottoman-era bridge in Mostar). Failed EC/U peace efforts led to a string of United Nations (UN) attempts with similar results. UN peacekeeping forces proved ineffectual in stemming the fighting or the atrocities, but a Muslim-Croat federative alliance (1994) helped stabilize the anti-Serb military position. In May 1995 Croatia, rearmed with United States support, launched a successful attack against Serbian positions in Croatia. Threatened with military collapse and NATO air intervention, and pressured by Russian diplomacy, Milošević reined in his Bosnian Serb surrogates and signed an American-brokered peace accord for the Bosnian war in Dayton, Ohio (November 1995), tentatively preserving a Muslim-Croat and Bosnian Serb federated state of Bosnia-Hercegovina.

CONQUESTS IN SLAVONIA AND BOSNIA, 1991–1994

━━━ Bosnian border
░ Croatian holdings
▫ Bosnian (Muslim) govt. holdings
▓ Serbian holdings

THE "DAYTON ACCORD" SETTLEMENT, 1995

━━━ Bosnian border
░ Muslim-Croat Federation
▫ Serbian Republic
▓ U. N. presence in eastern Slavonia

Map 52: The Kosovo Crisis, 1999

In 1966 Tito granted Serbia's autonomous provinces equal voting privileges with republics at the federal level, resulting in greater Albanian participation in Kosovo's provincial administration. Until 1966, the province's Albanian majority (90 percent of the inhabitants) lived under the discriminatory administrative dominance of Serbian authorities. The new Communist Albanian provincial representatives acted in a retaliatory fashion toward the resident Serbian minority, and also began calling for Kosovo's elevation to republic status. The post-Tito Yugoslav collective federal presidency refused to accede to the Kosovar Albanians' desires, since doing so would have violated Yugoslavia's constitutional foundations. (Constitutionally, only Yugoslav "nationalities" could possess republics; since an Albanian nation-state already existed outside of Yugoslavia, the Kosovar Albanians technically were a Yugoslav "national minority" and so were disqualified from having a republic.)

Albanian nationalist ideals took hold among university students, disaffected intellectuals, and workers in Kosovo, and in 1981 they staged demonstrations and riots. A third of the JNA was deployed in Kosovo to curb the disturbances and national paranoia grew among the Kosovar Serb minority. Exaggerated reports of Albanian anti-Serb atrocities appeared in the Serbian media and nationalist xenophobia. The paranoia in Serbia over Kosovo exploded in 1989 during a Serbian celebration of the 600th anniversary of the battle at Kosovo Polje. Slobodan Milošević's public expression of Serbia's support for the Kosovar Serbian nationalists at that event reopened old issues stemming from Yugoslavia's nature as a unitary federation of separate, distinct "nations" that essentially was dominated by Serbs, and sparked Yugoslavia's disintegration. (See Map 51.)

Serbian leader Milošević reduced the autonomy of Serbia's Kosovo and Vojvodina provinces (1989), and units of the JNA were stationed in Kosovo to quash increasing Albanian nationalist riots and demonstrations, which occurred throughout 1990. Many leading Albanian intellectuals, officials, and administrators were arrested; the Serbian constitution was amended to eradicate the remaining vestiges of Kosovo's autonomy (1990); and new laws were passed attacking the Kosovar Albanians' civil rights, language, and culture. The Kosovar Albanians continued their calls for Kosovo's elevation to republic status and Serbian anti-Albanian repression intensified — Albanian property rights were restricted, and new employment laws expelled more than 80,000 Albanians from their jobs.

In September 1990 a group of Kosovar Albanian nationalists proclaimed an independent Kosovo republic. Within a year (September 1991) they conducted a referendum among fellow disaffected Albanians that won popular support for their program but increased the ire of the Serbian authorities. By the end of 1991, Yugoslavia dissolved into five separate states — Slovenia, Croatia, Bosnia-Hercegovina, Macedonia, and a truncated Yugoslavia consisting of Serbia and Montenegro. Amid the shock of the initial Serbo-Croat war in late 1991, Albania extended unilateral recognition to an independent sovereign Kosovo in support of the repressed Kosovar Albanian nationalists. In May 1992, a secret election was held throughout Kosovo to create a Kosovo republican government, and Ibrahim Rugova, leader of a political party of intellectuals, was elected president. Rugova's Gandhi-like political approach primarily emphasized the rejection of Serbia's continued governing legitimacy in Kosovo, nonviolent opposition to the Serbian authorities, and winning international support for the Kosovar Albanians' national cause.

Throughout the years of warfare in Bosnia (see Map 51), the Kosovar Albanians' situation deteriorated. Albanian land was confiscated and handed over to Bosnian Serb refugees, who were settled as Serb colonists in Kosovo. Increased numbers of Albanians were thrown out of work, expelled from the university, or arrested, gradually discrediting Rugova's government among the general Albanian population. His ineffectiveness was highlighted when NATO and EC/U authorities essentially (but mistakenly) embraced Milošević as the guarantor of Balkan peace at the Dayton meeting (late 1995) ending the Bosnian debacle.

Throughout 1996 and 1997, a small group of Kosovar Albanian nationalists broke with Rugova and created the guerilla Kosovë Liberation Army (KLA). They instigated an uprising in 1998 that resulted in intensive Serbian military operations and anti-Albanian atrocities in Kosovo. Bolstered by general Serbian nationalist sentiment, Milošević steadfastly refused to temper Serbian treatment of Kosovar Albanians in the face of growing expressions of international displeasure. By late 1998, JNA forces had the KLA on the run, Serbian ultranationalist police and paramilitaries stepped up their anti-Albanian campaign of atrocities, and thousands of Kosovar Albanians fled the province in terror, while Milošević played cat-and-mouse diplomatic games with United States and other Western representatives seeking to bring about an end to the violence.

As the JNA buildup in Kosovo continued in early 1999 and atrocity stories multiplied, a last-ditch effort to forge a solution to the situation was made at Western insistence in Rambouillet, France, but broke down over mutual Serbian and Kosovar Albanian intransigence. In March, the Serbs unleashed their military against the KLA, and an orgy of "ethnic cleansing" was initiated. Over a million Albanian refugees flooded into neighboring Albania and Macedonia. Led by the United States, NATO air forces pounded Serbia for over two months before Milošević conceded defeat and withdrew his forces from the province (June). NATO peacekeeping troops moved into Kosovo to protect the returning refugees and prevent future violence. Despite the NATO presence, Albanian anti-Serb reverse "ethnic cleansing" persisted, and no definitive solution to the province's national problems was reached.

YUGOSLAVIA

Niš

Raška

SERBIA

Kuršumilja

Leskovac

Novi Pazar

Leposavić

MONTENEGRO

I

Ibar R.

Ibar R.

Rošaj

Mitrovica

Podujevo

Sitnica R.

Peć

II

Priština

Plav

Kosovo
Polje

Novo Brdo

Gračanica

Beli Drin R.

Dečani

Kamenica

Vranja

III

Dakovica

V

Gnjilane

Morava R.

IV

Bajram
Curri

Preševo

Drin R.

Prizren

Kačanik

Kumanovo

Kukës

Blače

ALBANIA

Skopje

Tetevo

Vardar R.

Drin i Zi R.

Gostivar

MACEDONIA

Veles

Morava R.

Border of Kosovo province, Serbia

NATO peacekeeping sector boundaries

NATO peacekeeping sectors

I   French   IV German
II Italian   V   American
III British

MILES

0   10   20

0   10   20

KILOMETERS

Selected Bibliography

The following list of sources is not comprehensive. For the most part, the works are general studies or commonly available atlases. Some have been included because their usefulness is broader than any single topic or because it is believed that they are of intrinsic interest to the general reader or student. The envisioned audience for this concise atlas is assumed to be English-speaking general readers and students, thus only works written in English are listed. While this might seem arbitrarily limited in scope, it is not. Those who possess the ability to read non-English languages will find plenty of additional titles to explore in the bibliographies and notes to many of the works cited.

Barker, Elisabeth. *Macedonia: Its Place in Balkan Power Politics.* London: Royal Institute of International Affairs, 1950.

Barraclough, Geoffrey, ed. *The Times Atlas of World History.* Revised ed. London: Times Books, 1979.

Bideleux, Robert, and Ian Jeffries. *A History of Eastern Europe: Crisis and Change.* London: Routledge, 1998.

Borsody, Stephen. *The New Central Europe: Triumphs and Tragedies.* Rev. and exp. ed. Boulder, CO: East European Monographs, 1993.

Cambridge Medieval History. Vol 1: Map supplement. Cambridge: Cambridge University Press, 1911.

Clogg, Richard. *A Concise History of Greece.* Cambridge: Cambridge University Press, 1992.

Crampton, Richard J. *A Concise History of Bulgaria.* Cambridge: Cambridge University Press, 1997.

———. *Eastern Europe in the Twentieth Century—And After.* 2nd ed. London: Routledge, 1997.

Crampton, Richard J., and Ben Crampton. *Atlas of Eastern Europe in the Twentieth Century.* London: Routledge, 1996.

Darby, H.C., and Harold Fullard. *The New Cambridge Modern History.* Vol. 14: *Atlas.* Cambridge: Cambridge University Press, 1978.

Davies, Norman. *Heart of Europe: A Short History of Poland.* Oxford: Oxford University Press, 1986.

Dvornik, Francis. *The Slavs in European History and Civilization.* New Brunswick, NJ: Rutgers University Press, 1962.

Eterovich, Francis H., and Christopher Spalatin, eds. *Croatia: Land, People and Culture.* 2 vols. Toronto: University of Toronto Press, 1964-70.

Fine, John V.A., Jr. *The Early Medieval Balkans: A Critical Survey From the Sixth to the Late Twelfth Century.* Ann Arbor: University of Michigan Press, 1983.

———. *The Late Medieval Balkans: A Critical Survey from the Late Twelfth Century to the Ottoman Conquest.* Ann Arbor: University of Michigan Press, 1987.

Goldstein, Ivo. *Croatia: A History.* Translated by Nikolina Jovanovic. Montreal: McGill-Queen's University Press, 1999.

Guldescu, Stanko. *The Croatian-Slavonian Kingdom, 1526-1792.* The Hague: Mouton, 1970.

———. *History of Medieval Croatia.* The Hague: Mouton, 1964.

Halecki, Oskar. *A History of Poland.* Translated by Monica M. Gardner and Mary Corbridge-Patkaniowska. New York: Roy, 1943.

———. *The Limits and Divisions of European History.* Notre Dame, IN: University of Notre Dame Press, 1962.

Held, Joseph, ed. *The Columbia History of Eastern Europe in the Twentieth Century.* New York: Columbia University Press, 1993.

Hupchick, Dennis P. *The Balkans: From Constantinople to Communism.* New York: Palgrave, 2002.

———. *Conflict and Chaos in Eastern Europe.* New York: St. Martin's Press, 1995.

———. *Culture and History in Eastern Europe.* New York: St. Martin's Press, 1994.

Hupchick, Dennis P., and Harold E. Cox. *The Palgrave Concise Historical Atlas of the Balkans.* New York: Palgrave, 2001.

Inalcik, Halil. *The Ottoman Empire: The Classical Age, 1300-1600.* Translated by Norman Itzkowitz and Colin Imber. London: Weidenfeld & Nicolson, 1973.

Itzkowitz, Norman. *Ottoman Empire and Islamic Tradition.* Chicago, IL: University of Chicago Press, 1980.

Jászi, Oscar. *The Dissolution of the Habsburg Monarchy.* Chicago, IL: University of Chicago Press, 1961.

Jelavich, Barbara. *History of the Balkans.* 2 vols. Cambridge: Cambridge University Press, 1985.

Jelavich, Charles, and Barbara Jelavich. *The Establishment of the Balkan National States, 1804-1920.* Seattle: University of Washington Press, 1977.

Judah, Tim. *The Serbs: History, Myth and the Destruction of Yugoslavia.* New Haven, CT: Yale University Press, 1997.

Kann, Robert A. *A History of the Habsburg Empire, 1526-1918.* Berkeley: University of California Press, 1974.

Kann, Robert A., and Zdenek V. David. *The Peoples of the Eastern Habsburg Lands, 1526-1918.* Seattle: University of Washington Press, 1984.

Kirschbaum, Stanislav J. *A History of Slovakia: The Struggle for Survival.* New York: St. Martin's Press, 1994.

Kolarz, Walter. *Myths and Realities in Eastern Europe.* London: Lindsay Drummond, 1946.

Korbel, Josef. *Twentieth Century Czechoslovakia: The Meaning of Its History.* New York: Columbia University Press, 1977.

Lampe, John R. *Yugoslavia as History: Twice There Was a Country.* Cambridge: Cambridge University Press, 1996.

Lang, David M. *The Bulgarians: From Pagan Times to the Ottoman Conquest.* Boulder, CO: Westview Press, 1976.

Longworth, Philip. *The Making of Eastern Europe.* New York: St. Martin's Press, 1994.

Macartney, C.A. *Hungary: A Short History.* Chicago. IL: Aldine, 1962.

Macartney, C.A., and A.W. Palmer. *Independent Eastern Europe: A History.* London: Macmillan, 1966.

MacDermott, Mercia. *A History of Bulgaria, 1393-1885.* New York: Praeger, 1962.

Magocsi, Paul Robert, and Geoffrey J. Matthews. *Historical Atlas of East Central Europe.* Seattle: University of Washington Press, 1993.

Malcolm, Noel. *Bosnia: A Short History.* New York: New York University Press, 1995.

———. *Kosovo: A Short History.* New York: New York University Press, 1998.

Matthew, Donald. *Atlas of Medieval Europe.* Oxford: Oxford University Press, 1983.

May, Arthur J. *The Habsburg Monarchy, 1867-1914.* Cambridge, MA: Harvard University Press, 1951.

McCarthy, Justin. *The Ottoman Turks: An Introductory History to 1923.* London: Longman, 1997.

Obolensky, Dimitri. *The Byzantine Commonwealth: Eastern Europe, 500-1453.* New York: Praeger, 1971.

Okey, Robin. *Eastern Europe 1740-1985: Feudalism to Communism.* 2nd ed. Minneapolis: University of Minnesota Press, 1986.

Osborne, R.H. *East-Central Europe: An Introductory Geography.* New York: Praeger, 1967.

Palmer, Alan. *The Decline and Fall of the Ottoman Empire.* New York: M. Evans, 1992.

———. *The Lands Between: A History of East-Central Europe Since the Congress of Vienna.* New York: Macmillan, 1970.

Palmer, R.R., ed. *Atlas of World History.* New York: Rand McNally, 1957.

Pavlowitch, Stevan K. *A History of the Balkans, 1804-1945.* London: Longman, 1999.

Petrovich, Michael B. *A History of Modern Serbia, 1804-1918.* 2 vols. New York: Harcourt Brace Jovanovich, 1976.

Pogonowski, Iwo Cyprian. *Poland: A Historical Atlas.* New York: Hippocrene, 1987.

Pollo, Stefanaq, and Arben Puto. *The History of Albania: From Its Origins to the Present Day.* London: Routledge & Kegan Paul, 1981.

Pounds, Norman J.G., and Robert C. Kingsbury. *An Atlas of European Affairs.* New York: Praeger, 1964.

Pribichevich, Stoyan. *Macedonia: Its People and History.* University Park: Pennsylvania State University Press, 1982.

Ramet, Sabrina P. *Nationalism and Federalism in Yugoslavia, 1962-1991.* 2nd ed. Bloomington: Indiana University Press, 1992.

Rónai, Andrew. *Atlas of Central Europe.* Budapest: Count Paul Teleki Research Institute, 1945.

Rothenberg, Gunther E. *The Military Border in Croatia, 1740-1881: A Study of an Imperial Institution.* Chicago, IL: University of Chicago Press, 1966.

Rothschild, Joseph. *East Central Europe Between the Two World Wars.* Seattle: University of Washington Press, 1974.

Rothschild, Joseph, and Nancy M. Wingfield. *Return to Diversity: A Political History of East Central Europe Since World War II.* 3rd ed. New York: Oxford University Press, 2000.

Runciman, Steven. *A History of the First Bulgarian Empire.* London: G. Bell, 1930.

Sedlar, Jean W. *East Central Europe in the Middle Ages, 1000-1500.* Seattle: University of Washington Press, 1993.

Seton-Watson, Hugh. *The East European Revolution.* 3rd ed. New York: Praeger, 1956.

Seton-Watson, R.W. *A History of the Roumanians.* New York: Shoe String Press, 1934.

Shaw, Stanford J., and Ezel K. Shaw. *History of the Ottoman Empire and Modern Turkey.* 2 vols. Cambridge: Cambridge University Press, 1976-77.

Shepherd, William R. *Shepherd's Historical Atlas.* 9th ed., rev. and updated. New York: Barnes & Noble, 1976.

Stavrianos, L.S. *The Balkans Since 1453.* New York: Holt, Rinehart & Winston, 1958.

Stoianovich, Traian. *Balkan Worlds: The First and Last Europe.* Armonk, NY: M.E. Sharpe, 1994.

Stokes, Gale. *Three Eras of Political Change in Eastern Europe.* New York: Oxford University Press, 1997.

———. *The Walls Came Tumbling Down: The Collapse of Communism in Eastern Europe.* New York: Oxford University Press, 1993.

Sugar, Peter F. *Southeastern Europe under Ottoman Rule, 1354-1804.* Seattle: University of Washington Press, 1977.

Sugar, Peter F., Péter Hanák, and Tibor Frank, eds. *A History of Hungary.* Bloomington: Indiana University Press, 1990.

Sugar, Peter F., and Ivo J. Lederer, eds. *Nationalism in Eastern Europe.* 2nd ed. Seattle: University of Washington Press, 1994.

Swain, Geoffrey, and Nigel Swain. *Eastern Europe Since 1945.* New York: St. Martin's Press, 1993.

Tapié, Victor-L. *The Rise and Fall of the Habsburg Monarchy.* Translated by Stephen Hardman. New York: Praeger, 1971.

Taylor, A.J.P. *The Habsburg Monarchy, 1809-1918: A History of the Austrian Empire and Austria-Hungary.* New York: Harper & Row, 1965.

Thomson, S. Harrison. *Czechoslovakia in European History.* Hamden, CT: Archon, 1965.

Vakalopoulos, Apostolos E. *The Greek Nation, 1453-1669: The Cultural and Economic Background of Modern Greek Society.* Translated by Ian Moles and Phania Moles. New Brunswick, NJ: Rutgers University Press, 1976.

―――. *History of Macedonia, 1354-1833.* Translated by Peter Megann. Thessaloniki: Institute for Balkan Studies, 1973.

―――. *Origins of the Greek Nation: The Byzantine Period, 1204-1461.* Translated by Ian Moles. New Brunswick, NJ: Rutgers University Press, 1970.

Vickers, Miranda. *Albania: A Modern History.* London: I.B. Tauris, 1994.

Vidal-Naquet, Pierre, and Jacques Bertin. *The Harper Atlas of World History.* Translated by Chris Turner, et al. New York: Harper & Row, 1987.

Vucinich, Wayne S. *The Ottoman Empire: Its Record and Legacy.* Reprint ed. Huntington, NY: Krieger, 1979.

Walters, E. Garrison. *The Other Europe: Eastern Europe to 1945.* Syracuse, NY: Syracuse University Press, 1988.

Wandycz, Piotr S. *The Price of Freedom: A History of East Central Europe from the Middle Ages to the Present.* London: Routledge, 1992.

Wilkinson, H.R. *Maps and Politics: A Review of the Ethnographic Cartography of Macedonia.* Liverpool: University Press of Liverpool, 1951.

Williams, Joseph E., ed. *World Atlas.* Englewood Cliffs, NJ: Prentice-Hall, 1960.

Wolff, Robert L. *The Balkans in Our Time.* Cambridge, MA: Harvard University Press, 1956.

Woodhouse, Christopher M. *Modern Greece: A Short History.* London: Faber, 1977.

World History Atlas. Rev. ed. Maplewood, NJ: Hammond, 1993.

Wynot, Edward D. *Cauldron of Conflict: Eastern Europe, 1918-1945.* Wheeling, IL: Harlan Davidon, 1999.

Index

This atlas is indexed according to map number. Non-italicized numbers refer to text pages; italicized numbers refer to map pages.